FAR INFRARED ASTRONOMY

OTHER TITLES OF INTEREST

FAR INFRARED ASTRONOMY

edited by

MICHAEL ROWAN-ROBINSON

Department of Applied Mathematics
Queen Mary College, London E.1.

PROCEEDINGS OF A CONFERENCE HELD AT
CUMBERLAND LODGE, WINDSOR, U.K.
ON JULY 11th-13th, 1975

SPONSORED BY THE ROYAL ASTRONOMICAL SOCIETY

PERGAMON PRESS

OXFORD · NEW YORK · TORONTO
SYDNEY · PARIS · FRANKFURT

U. K.	Pergamon Press Ltd., Headington Hill Hall, Oxford OX3 0BW, England
U.S.A.	Pergamon Press Inc., Maxwell House, Fairview Park, Elmsford, New York 10523, U.S.A.
CANADA	Pergamon of Canada, Ltd., P.O. Box 9600, Don Mills M3C 2T9, Ontario, Canada
AUSTRALIA	Pergamon Press (Aust.) Pty. Ltd., 19a Boundary Street, Rushcutters Bay, N.S.W. 2011, Australia
FRANCE	Pergamon Press SARL, 24 rue des Ecoles, 75240 Paris, Cedex 05, France
WEST GERMANY	Pergamon Press GmbH, 6242, Kronberg-Taunus, Pferdstrasse 1, Frankfurt-am-Main, West Germany

First edition 1976

Library of Congress Cataloging in Publication Data

Main entry under title:

Far infrared astronomy.

"Supplement to Vistas in astronomy."
Includes indexes.
1. Infra-red astronomy-Congresses. I. Rowan-Robinson, Michael.
II. Royal Astronomical Society. III. Vistas in astronomy.
QB470.A1F37 1976 522'.6 75-42492
ISBN 0-08-020513-5-Y
0-08-020591-7-R

Printed in Great Britain by A. Wheaton & Co., Exeter

0 08 020513 5 Y
0 08 020591 7 R

CONTENTS

Contents

PART 6:   THEORETICAL MODELS

PREFACE

Until recent years the far infrared (IR) has been the last great gap in the
electromagnetic spectrum available to the astronomer. This book, based on the
first international conference devoted to the field of far IR astronomy as a
whole, held at Windsor, on July 9th - 11th, 1975, shows the different ways
this gap has been bridged. I hope this collection of papers will provide a
stimulating introduction to this rapidly growing field, which stands roughly
where X-ray astronomy stood 5 years ago and where radio-astronomy did 25 years
ago.

Part 1, Instrumentation & Atmospheric Constraints, gives an idea of the
difficulty of far infrared observations. Traub's calculations of atmospheric
transmission at different altitudes illustrate how, apart from the 350 μm
window available to good high altitude sites, balloon, aircraft or satellite-
borne telescopes are necessary for observations at wavelengths between 20 μm
and 1 mm. Furniss et al and Fazio et al describe balloon-borne telescopes,
and Kleinmann compares their IR performance with that of the planned
satellite-borne Large Space Telescope. As Sollner shows, atmospheric noise
can be greatly reduced by differential sky-chopping, for which a new technique
is described by Lemke et al. Hofmann et al and Traub describe designs for
high resolution interferometers to observe far IR emission lines. The
highlights of Part 2, Solar and Jovian Atmospheres, are the balloon-borne
Michelson interferometer spectra of Jupiter by Furniss et al and the
discussion by Beckman & Ross of their solar scans made from Concorde during a
solar eclipse.

One of the most important areas of far IR astronomy is the Cosmic Microwave
Background, the subject of Part 3, since the peak intensity of a 2.7 K
blackbody occurs near 1 mm. Measurements of the spectrum of the background,
like those of the QMC group described by Robson, and planned by the Leeds
group (Mercer et al), will help to decide whether the radiation is indeed the
relic of the fireball phase of a big bang universe, or is due to some more
local source (for example, integrated light from galaxies thermalised by large
graphite whiskers, as proposed by Narlikar et al). One of the difficulties of
these experiments is that even at balloon altitudes the emission from the

residual atmosphere is considerable and, as can be seen from the general
discussion on the background, the subject of controversy.

Line Astronomy has yet to penetrate the 20 μm − 1 mm band, so the papers of
Part 4 deal mainly with molecular line observations of HII regions with
radio techniques (H_2CO by Gardner & Whiteoak, H_2O by Little, CO by Gillespie
& Phillips). These show the intimate relationship of molecules to dust, as
revealed by far IR continuum emission. Anderegg et al also describe a ground-
based search for emission lines in the 10 and 20 μm bands with a very high
resolution Michelson interferometer designed for use in the far IR on NASA's
airborne (C141) observatory.

In Part 5, Continuum Emission, Shivanandan et al and Turon et al report the
results of airborne multiband photometer observations of HII regions, Clegg
et al describe ground-based millimetre observations of galactic and extragalactic sources. Emerson & Jennings use broad-band far IR observations of
HII regions to deduce the location and composition of the dust which is
absorbing much of the light from the ionizing stars. Mezger reviews radio
and IR data on the W3 region and gives a model both of the structure of dusty
HII regions and of their possible evolution. Further discussion of models of
dust clouds radiating in the far IR is to be found in Part 6, Theoretical
Models, in the papers by Andriesse, Aanestad and Edmunds & Wickramasinghe.
The nature of the grains remains a matter of controversy, ranging from ice or
silicates to formaldehyde polymers (Cooke & Wickramasinghe). Hong & Greenberg
discuss the effect of very small grains on the radiation field and Silk
produces a theory of the formation, growth and destruction of grains in the
protostellar environment.

John Bastin, who organised the conference, provides an introduction. My warm
thanks to Jane Percival and Lola Buer for persevering with the typing of the
book.

M.R.R.   Sept 1975

INTRODUCTION

by Professor J. Bastin
Department of Physics, Queen Mary College, London E1 4NS.

The desire on the part of some astronomers to establish an enclave for themselves is perhaps the least reason for the formation of infrared astronomy
as a field in its own right. The strong reasons for the separate identity
concern instrumental and observational technique; but it seems unlikely that
the subject matter of this new field will ever be distinct from other branches
of astronomy. Indeed this should not be its aim, for some of the most
interesting advances in general recent astronomy have come as the result of
the comparison of observations in widely differing wavelength ranges.

Until the last few years far infrared astronomical observation was entirely
ground based and it continues to be largely carried out in this mode. In the
wavelength range between 40μm and 200μm pure rotational absorption of
water vapour in the earth's atmosphere forbids ground based observation from
even the highest and driest sited observatories; and there is thus a natural
division between far infrared astronomy and the more traditional near infrared range below about 25μm.

At the longer wavelength bound the demarcation between far infrared astronomy
and high frequency radio astronomy is more complex. At wavelengths below about
3 mm ($\nu = 10^{11}$ Hz) atmospheric absorption is appreciable. Because of
pressure broadening there is absorption throughout the windows between adjacent
strong lines. This absorption becomes progressively greater at higher frequu-
encies, and there are distinct advantages in the choice of dry high altitude
sites for far infrared observations above 1 mm, whilst below this wavelength
such a choice is essential. For this reason telescopes for far infrared
observation cannot normally be mounted at observatories used for the more
conventional radio astronomy. Until recently this dichotomy at about 3 mm
was accentuated by two matters of instrumental design - namely the construction
of telescopes primaries and the choice of detectors.

A diagram showing the angular resolution (reciprocal of the smallest resolvable
angle) of the largest available telescopes from the far ultraviolet, to the
longest wavelengths used in radio astronomy, shows a most noticeable dip in

the centre of the far infrared wavelength range. This is in large part a
consequence of the limitations of engineering methods. To be a useful
receiver at a given wavelength a parabolic collector must be perfect in
profile to a tolerance of about a sixteenth of this wavelength. Nearly four
centuries of work have produced visual telescopes of angular resolution
approaching 10^7. In the radio region similar resolution is obtainable only
by the use of spatial interferometry, but in the millimetre wavelength region
such interferometric techniques present great difficulty and the use of
parabolic collectors presents the only method currently available. The
profile accuracy of such collectors made by machining methods is such that
they have only been used with high efficiency at wavelengths above about
3 mm. Below this wavelength at least one such telescope has been used at
greatly reduced efficiency but in most cases recourse to optically polished
conventional telescopes has been necessary.

With detectors there is a similar transition of method at low millimetric
wavelengths. Heterodyne receivers, the standard detector for short wave
radio astronomy, are largely replaced in the far infrared region by cooled
bolometers and photoconductors. Both types of detector are about as
efficient in the detection of a broad spectral source but the heterodyne
receiver is finely tunable and its high sensitivity over a narrow wavelength
range make it the only feasible instrument for investigation of molecular
transitions. Recently the use of such detectors has been extended with
great advantage for astronomical observations at wavelengths only just above
1 mm. There are great incentives in such work and in spite of the technical
problems the development to submillimetric wavelengths may well prove most
significant in the development of far infrared astronomy in the next decade.
It seems likely also that there will be sufficient incentive to overcome
the engineering problems of building parabolic telescopes in excess of 10 m
aperture and of profile tolerance suitable for submillimetre reception.

The short wavelength bound of far infrared astronomy is also now being eroded,
in this case by the advent of rocket, balloon and aircraft based observations.
The water vapour scale height of the atmosphere is about 2 km, compared with
8 km for the principal atmospheric gases. By flying a balloon at about
30 km, that is at a height above all except a few percent of the major
constituents, it is therefore possible to be above all except a few thousands

of a percent of the water vapour in the atmospheres. At this height other
constituents of the high atmosphere (principally ozone) cause absorption
of about the same magnitude but in most cases their effect is small and can
be taken into account in the analysis of measurements.

The divisions which a decade ago made far infrared astronomy a self contained
entity are thus now largely blurred. Nevertheless much observational work
will continue for many years to be carried out from high altitude sites
between 2 and 5 km above sea level, and this is sufficient to preserve
the identity of the subject. Indeed this tendency may well be accentuated,
for several telescopes (very large) of optical quality are now being
constructed for installation at high sites.

As in the case of main sequence stars of surface temperatures of thousands of
degrees Kelvin, with a spectrum maximising within or near the visual wavelength range ($\lambda \sim hc/KT$) so we would expect the thermal sources prominent in
far infrared astronomy to have temperatures of a few or a few tens of degrees
Kelvin. (The cosmic background radiation, with the interesting part of its
spectrum in this range, is one such source).

We might expect that clouds of dust and gas should also present such sources.
What has perhaps come as a surprise is the large angular extent of the regions
over which a measurable signal is at present detectable. Indeed a future
problem may well concern the determination of the magnitudes of the components
of intensity assignable to a number of physically distinct sources which
overlap when projected on the celestial sphere.

Of course for a source to emit thermally it cannot be transparent. Dust
grains certainly satisfy the opacity condition for emission but the optical
thickness of many gas clouds in the far infrared is often much less than
unity. The various molecular transitions thus show up as bright lines
(cf the chromospheric lines seen during an eclipse) and there seems little
doubt that line astronomy will continue to be extended until it covers the
whole far infrared range.

And for the future there will doubtless be many other experimental developments.

the greater utilisation of Fourier transform interferometry, the employment of techniques to detect polarisation of far infrared radiation and the use of satellite-based equipment.

Finally, in our justifiable fascination with far infrared radiation from interstellar matter in this and other galaxies we have perhaps lost sight of the possibility of observation within our own solar system. At present it is not impossible to detect an object giving one flux unit (1 f.u. = 1 Jansky = 10^{-26} W m^{-2} Hz^{-1}) of radiation at 1 mm: yet from several of the planets we receive many thousands of times this flux. As the number of far infrared astronomers grow it may be hoped that at least some may decide to specialise in this field.

PART 1

INSTRUMENTATION AND ATMOSPHERIC
CONSTRAINTS

BALLOON-BORNE TRANSFORM SPECTROSCOPY

W. A. Traub

Center for Astrophysics, Cambridge, Mass. 02138, U.S.A.

Abstract The design and construction of a high-resolution far
infrared Fourier transform spectrometer for use on the
Smithsonian balloon-borne 1-metre telescope is described. The
instrument will operate at a resolution of about 0.1 cm^{-1} in the
region 25 to 150 μm, and will be used to obtain spectra of
Jupiter, Venus, Orion and other HII and molecular cloud regions,
and the terrestrial stratosphere.

Introduction

In order to increase our understanding of the physical conditions in
planetary atmospheres, HII regions and galactic molecular clouds, we expect
to begin construction soon on a far infrared Fourier transform spectrometer
which will be used with the 102-cm balloon-borne telescope of the Centre for
Astrophysics. A portion of our preliminary design effort has gone into a
calculation of the expected infrared absorption spectrum of the terrestrial
atmosphere. I will first briefly describe the results of this calculation,
and then discuss our preliminary design for a spectrometer.

Atmospheric Transmission

There are a number of situations in which it is useful to have some knowledge
of the far infrared transmission characteristics of the terrestrial
atmosphere. Since experimental data is often lacking, a theoretical approach
is necessary; one particular advantage in doing such a calculation is that it
is easy to compare spectra with various altitudes and compositions.

The basic source of information here is the atmospheric absorption line
parameter tape generated at the U.S. Air Force Cambridge Research Labs.
(McClatchey et al., 1973). This compilation lists the wavenumber, line

strength, pressure-broadening coefficient and energy level of the lower
state for over 109,000 known transitions of H_2O O_3, O_2, CO_2, CO, N_2O and
CH_4 between about 0.76μ and $3.26mm$ (some lines of O_2 at $\lambda \sim 1mm$ are not on
the tape). The molecular abundances are listed in Table 1.

TABLE 1

Molecular abundances, effective pressures and temperatures as used in the
Curtis-Godson approximation. The H_2O abundances in the last three columns
correspond to 2.25, 0.26 and 0.040 precipitable microns respectively; the
H_2O at 4.2 km is assumed to have a scale height of 1.85 km. Abundances are
adapted from Farmer (1974) (and references therein), Morrison et al.(1973),
Patel (1974), and U.S. Standard Atmosphere (1962). The abundances listed
are for unit air mass; an additional factor of 2 is included in the actual
calculations corresponding to a zenith angle of 60°.

	4.2 km (Mauna Kea)	14 km (Aircraft)	28 km (Balloon)	41 km (Balloon)
O_2	209460.0 ppmv	209460.0	209460.0	209460.0
CO_2	325.0 ppmv	325.0	325.0	325.0
CH_4	1.5 ppmv	1.1	0.8	0.4
N_2O	0.25 ppmv	0.20	0.20	0.20
CO	0.07 ppmv	0.06	0.06	0.06
H_2O	1200 micron	2.5 ppmv	2.5 ppmv	2.5 ppmv
O_3	7.28 E18 cm^{-2}	6.37 E18	1.85 E18	1.70 E17
P	600.0 mbar	141.6	16.2	2.52
P(eff)	300.0 mbar	70.8	8.10	1.26
P(H_2O)	506.0 mbar	70.8	8.10	1.26
P(O_3)	36.4 mbar	30.2	7.09	1.84
T(eff)	228.0 K	217.0	230.0	268.0

It is assumed that each species is well mixed above the observer's altitude,
except for H_2O and O_3, where in general different concentration profiles are
assigned. We use the Curtis-Godson approximation (Goody, 1964), so that the
real exponential atmosphere is replaced by an equivalent layer having a
thickness of one scale height, a pressure equal to one-half the base pressure
(P/2), and a temperature taken to be that which occurs in the real atmosphere

at a pressure level of P/2. The temperature-pressure profile is a year-round average taken from the U.S.Standard Atmosphere, 1962; the O_3 profile comes from this reference also, but it has been normalized to correspond to a sea-level abundance of 0.28 cm-atm.

Line profiles are calculated by assuming that both pressure broadening ($\Delta\sigma_L$(HWHM) = α_0 P(eff)/P_0) and Doppler broadening ($\Delta\sigma_G$ (1/e half-width) = 4.30 × $10^{-7}\sigma$(T/M$^{1/2}$) are present, so that in general the line profile is a Voigt function with Voigt parameter A = $\Delta\sigma_L/\Delta\sigma_G$. For an exponential, isothermal atmosphere the line profile is a logarithmic function which rapidly reduces to the Lorentz profile in the wing; there is no simple form if the Doppler broadening is included. In the region defined by A < 2.0 and ($\sigma-\sigma_0$) < 12.0 $\Delta\sigma_G$ the Voigt function is used; otherwise it is adequate to use a simple Lorentz profile since the error in the resultant spectra will be of the order of 1 percent or less. The calculating grid has step sizes which are comparable to or larger than a typical line width (grid = 0.050cm^{-1} at 4.2 km, 0.010cm^{-1} at 14 km, and 0.01cm^{-1} at 28 and 41 km). At the centre point of each line the average optical depth is computed over a single grid element; elsewhere the line strength across a grid element is approximated by its value at the centre. Line wings are calculated out to an optical depth of 0.0005. The calculated transmission spectra are averaged over intervals of 0.05cm^{-1} before plotting. Only those lines which will yield a central optical depth of more than 0.0005 on the averaged spectra are included. The line of sight is taken to be 60° from the zenith, so that the air mass is 2.0. Spectra are calculated in units of 10 to 100cm^{-1}, depending upon the grid spacing; the absorption in each such unit also includes the wings from lines located from 1 to as much as 10cm^{-1} away. Calculated spectra from 25 μm to 1000 μm for the altitudes of interest are shown in Figs 1 to 4.

The broad-band absorption characteristics of the atmosphere were derived from the high-resolution calculations by averaging the calculated transmission spectra over units of 10cm^{-1} (or $\Delta\sigma$ = 100cm^{-1} for σ > 2000cm^{-1}); the results are shown in Fig. 5, where both transmission (T) and emissivity (1-T) are displayed. The rapid increase of transmission with increasing altitude is caused in part by the drop in H_2O mixing ratio and also in part by the decreased Lorentz line strength (in the wings). It is clear that photometry and especially spectroscopy should enjoy substantial benefits at balloon altitudes as compared to aircraft altitudes.

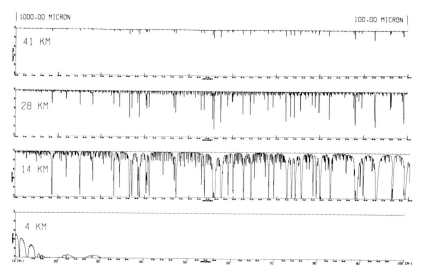

<u>Fig. 1</u> Atmospheric transmission at a resolution of 0.05 cm^{-1} from 1000 μm to 100 μm.

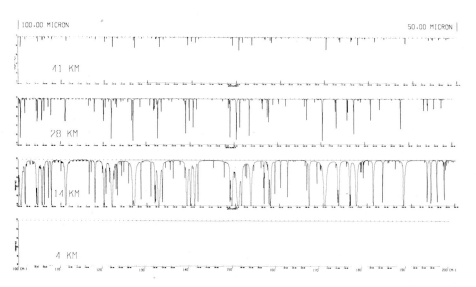

<u>Fig. 2</u> Atmospheric transmission at a resolution of 0.05 cm^{-1} from 100 μm to 50 μm.

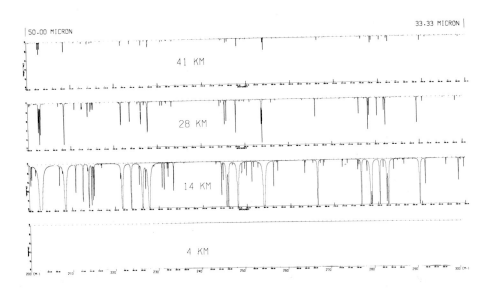

Fig. 3 Atmospheric transmission at a resolution of 0.05 cm^{-1} from 50 μm to 33 μm.

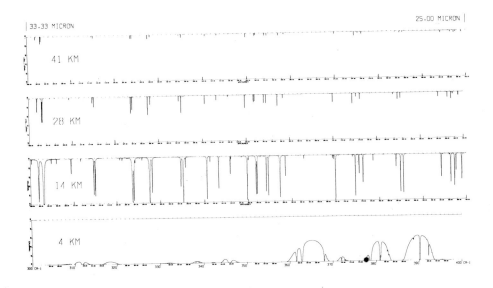

Fig. 4 Atmospheric transmission at a resolution of 0.05 cm^{-1} from 33 μm to 25 μm.

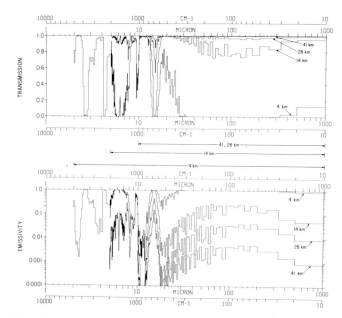

Fig. 5 Atmospheric transmission and emissivity at a resolution of
10 cm⁻¹ for σ < 2000cm⁻¹ and 100 cm⁻¹ for σ > 2000 cm, for the
abundances given in Table 1 and an air mass factor of 2.0. The
spectral range is extended to shorter wavelengths for the lower
altitudes, as indicated.

Fourier Transform Spectrometer

For far infrared spectroscopy, it is clear from the above results that the
atmosphere is much more transparent at balloon altitudes than it is at
aircraft altitudes. Using our 102-cm balloon-borne telescope, we intend to
take advantage of this increased transparency (and reduced emissivity) by
using a Michelson type of Fourier transform spectrometer, with a resolution
of about 0.1 cm^{-1} (apodized) or 0.05 cm^{-1} (unapodized).

A Michelson spectrometer offers the well-known advantages of multiplexing and
large etendue (or throughput), both of which are important in the far infrared
where the detectors are extremely noisy and sources subtend relatively large
angles. We expect to begin construction soon on an instrument which possesses
the preliminary design characteristics discussed in this section; a possible
configuration is shown in Fig. 6.

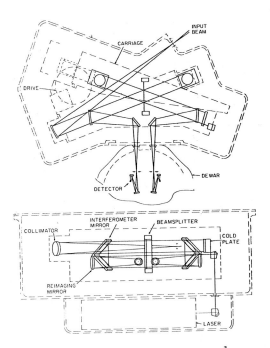

Fig. 6   Preliminary design for a 0.1 cm^{-1} Michelson
spectrometer to operate primarily from 25 μm to 150 μm. The main
interferometer mirrors can be corner reflectors (as shown),
cube-corner reflectors, or cat's eyes. The cold plate will likely
be replaced by a series of relay mirrors that view the cold dewar
interior through a window.

The main input beam will have a diameter of about 90" arc or 6 mm; the second
input beam will view a cold surface designed such that the central
interference fringe will be substantially reduced in amplitude compared to
that expected from a single-input Michelson. The recorded spectra will thus
be the complete emission and absorption composite from the source, sky,
telescope and instrument. By aiming the telescope at nearby "blank" sky, we
get a comparison spectrum which can then be subtracted later in the computer.
This avoids the problem of achieving a balanced output with a chopping
secondary; also we can complete an interferogram relatively rapidly (∿30 sec),
whereas with a chopping secondary at 20 Hz it would take ∿ 30 min. to complete
one scan.

There are two output beams going to a pair of germanium (three-part) bolometers, so that if one detector fails, we will still have the other. The sum

of the outputs can in principle be used to normalize signal fluctuations caused by telescope pointing errors. A DC monitor on each channel will provide low-frequency intensity variation information, while the interferogram will be AC coupled.

The moving mirror will be driven about 5 cm, so the apodized resolution is about 0.1 cm^{-1}; for many experiments it should be possible to operate in an unapodized mode and the resolution will then approach 0.05 cm^{-1}, which is the same as the averaging interval used in Figs. 1-4. To eliminate the necessity for extremely accurate translational parallelism in the mirror drive, we will likely use cat's eye retroreflectors. The rms fluctuation in mirror velocity will be held to about 0.1%, and the sampling interval rms error should be within 60 Å. To achieve this we will use a velocity feedback servo loop operating on an input signal derived from the interference fringes of a 6328 Å beam from a Hewlett-Packard Zeeman-split He-Ne laser. The instrumental sensitivity is expected to be of the order of 1×10^{-12} W m^{-2} in a one-second integration with 0.1 cm^{-1} resolution and signal-to-noise of unity. This will be due in about equal parts to detector phonon noise and signal shot noise.

The spectral range (at $\geq$ 50% relative efficiency) will initially be about 25 μm to 150 μm, limited by the mylar beamsplitter. The instrument will be able to be adapted for operation in the mid-infrared (λ < 25 μm) from the ground or on aircraft platforms if desired.

The instrument will be pre-cooled before launch to about 220 K and evacuated to about 10 torr. This gives us a reduced background signal, helps ensure alignment stability, and helps eliminate acoustic coupling of vibrations to the beamsplitter. Insulation will keep out the heat of the lower atmosphere and skin heaters will eliminate icing. Mylar windows will likely be used to interface to the outside, and sapphire filters will be used in the dewars.

An auxiliary offset-guiding star tracker will provide an error signal to the gyros and should be able to yield tracking to several arcsec over many minutes of time.

Acknowledgments

Others involved in this work are N.P. Carleton, G.G. Fazio, R.M. Goody, and R. W. Noyes; also M.T. Stier contributed a great deal to the atmospheric transmission calculation. A complete description of the atmospheric spectra will be published elsewhere. This work is supported in part by NASA grant NGR-09-015-047.

References

Goody, R.M., 1964, Atmospheric Radiation, I. Clarendon Press.
Farmer, C.B., 1974, Can.J.Chem. 52, 1544.
McClatchey, R.A., Benedict, W.S., Clough, S.A., Burch, D.E., Calfee, R. F.,
    Fox, K., Rothman, L.S., and Garing, J.S., 1973, AFCRL Atmospheric
    Absorption Line Parameters Compilation, AFCRL-TR-73-0096.
Morrison, D., Murphy, R.E., Cruikshank, D.P., Sinton, W.M., and Martin, T.Z.,
    1973, Pub.Astron.Soc.Pacific, 85, 255.
Patel, C.K.N., Burkhardt, E.G., and Lambert, C.A., 1974, Science, 184, 1173.
U.S. Standard Atmosphere, 1962, NASA, USAF, USWB, U.S. Government Printing
                                          Office, Washington, D.C.

DISCUSSION

Moorwood What is your philosophy on sky chopping?

Traub We intend not to use a sky chopping system, but instead to simply record the AC part of the interferogram at both output ports from the interferometer. In principle, one can use the sum of these signals to normalize either channel against scintillation or pointing errors. There is also the advantage that sky spectra are explicitly recorded, so that one can search for stratospheric emission lines from chemically interesting molecular species, and further, the absorption of astronomical spectra by atmospheric lines can be inferred. Finally, secondary chopping forces you to record the interferogram at a relatively slow speed (about 30 minutes in our case), but since one cannot expect the guiding and sensitivity to remain constant over such long periods, it is preferable to be limited only by the detector response speed, giving a scan time of about 30 seconds in our instrument.

Welsh In your data analysis do you intend to use the standard Cooley-Tukey algorithm or the maximum entropy method of Ables J.G. (1974, Astr.Astrophys. Suppl. 15, 383)?

Traub The Cooley-Tukey algorithm.

Marchant Do you have any information on spatial and temporal variation in upper atmospheric emission?

Traub No, we do not have any data on this, but we do expect that the variations will be much smaller from a balloon platform than from an aircraft.

Joseph Have you examined possible difficulties in studying some of the astrophysically interesting lines at balloon altitudes due to the proximity of strong water vapour lines?

Traub We have not gone into detailed studies of possible observations, partly because the interstellar lines arise from fine structure levels which are themselves not too accurately known.

IMPLICATIONS OF ATMOSPHERIC FLUCTUATIONS FOR FAR INFRARED ASTRONOMY

T. C. L. G. Sollner

Queen Mary College, London E.1

Abstract   Observations of atmospheric emission and transmission fluctuations in the 350 µm window have been made under a variety of observing conditions. The amplitude of this source of noise is large compared to detector noise, and has a frequency dependent power spectrum of the form $P \propto f^{-n}$, $n = 0.8 \pm 0.2$. The effect of differential sky chopping is to decrease the observable emission noise by about a factor of ten, thus making it comparable to detector noise power.

Introduction

A primary source of signal obscuration in many types of physical experiments is the fluctuation of some unavoidable "DC" background level. This is the situation in far infrared astronomy, in which a large background is encountered from the atmosphere, and in some cases from the telescope elements themselves. We have examined the fluctuations of atmospheric emission, due primarily to water vapour, in the spectral region between $20 - 30$ cm^{-1} under conditions identical to those used for astronomical observation in this atmospheric window. This work will be published in detail elsewhere, so only an outline of the apparatus and results will be presented here.

Equipment

The 24-in Cassegrain telescope administered by the University of Denver was used for these measurements. It is located at an elevation of 13,800 ft on the top of Mt.Evans, Colorado. The detector was a liquid helium cooled germanium bolometer. A focal plane chopper alternately obscured one of two apertures 5' arc in diameter separated by 6' arc, at a frequency of 90 Hz. One aperture could be closed to operate in the single beam mode. A cold low-pass filter was used to limit detector response to less than 40 cm^{-1} and an ambient temperature mesh interference filter further restricted the bandpass to 20-30 cm^{-1}.

Figure 1 shows the electronics used to take and record the data. The signal was averaged for one second every two seconds and recorded. Low pass filtering was used to ensure against aliasing.

Results

The sampled sky signal obtained from a typical dual beam observation is shown in Fig. 2. The power spectrum of this signal appears in Fig. 3. The straight line is a least-squares fit to the data. From about 30 spectra of this type, taken under varying conditions, the following conclusions can be drawn:

1. In the single beam mode, the sky emission noise power is an order of magnitude greater than detector noise power.

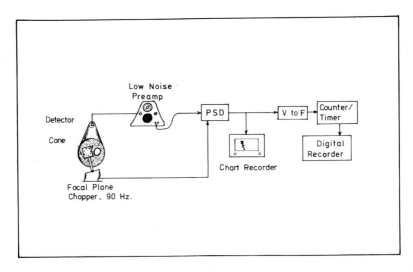

Fig. 1.   Electronics used to measure and record the atmospheric emission and transmission signal.

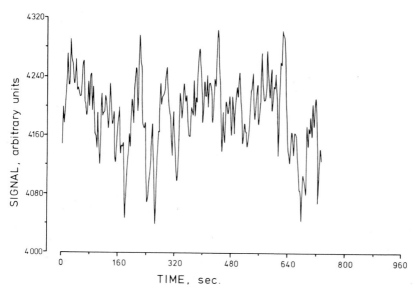

Fig. 2. Typical sky emission signal as a function of time measured in the differential chopping mode of operation.

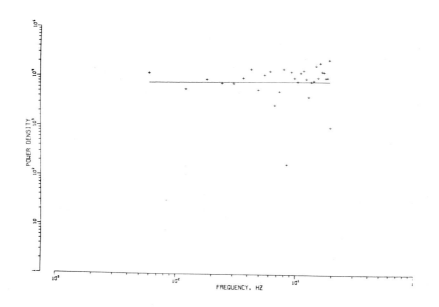

Fig. 3. Power spectrum of the signal shown in Fig.2. The solid line is a straight-line fit to the data. The vertical axis is in arbitrary units.

2. The power spectrum in both emission and transmission has a frequency dependence much steeper than that shown in Fig.3, of the form $P(f) \propto f^{-n}$, where $n = 0.8 \pm 0.2$, in the spectral region $0.6 - 20 \times 10^{-2}$ Hz.

3. Differential (two beam) chopping reduces the noise power from the atmosphere to approximate equivalence to detector noise ($\sim 10^{-12}$ W Hz$^{-\frac{1}{2}}$), and the spectrum becomes flat, resembling Fig. 3.

In view of these results, two beam chopping appears much superior to single beam operation. As detectors are improved, it will be possible to measure more accurately the frequency dependence of the power spectrum of the noise, when differentially chopping.

Acknowledgments

We would like to thank P.A.R. Ade, J.E. Beckman and E.I. Robson for assistance with the measurements, J.C.G. Lesurf for help in electronic design and construction, and P.E. Clegg for many helpful discussions. Also we express our appreciation to the University of Denver for observing time and for their assistance during that time.

SOME BALLOON-BORNE I R TELESCOPE DEVELOPMENTS AT UNIVERSITY COLLEGE LONDON

I. Furniss, R. E. Jennings, W. A. Towlson, T. E. Venis and B. Y. Welsh

Department of Physics and Astronomy, University College, London W.C.1

Abstract   The design of the balloon-borne 60-cm telescope system under construction at University College London is described, together with the modulation system to be used.

The 39 cm aperture telescope and stabilised platform which was developed by Tomlinson, Towlson and Venis (1974) and made in the Department of Physics and Astronomy is to be supplemented by a new and larger 60 cm aperture system. The present telescope has, to date, been flown on 16 occasions.

The new stabilised platform and telescope has been designed and developed by the Engineering Design Group in the Department and will be described later by Towlson and Venis. Briefly, this platform has two orthogonal fine control axes, elevation and cross elevation, to decouple the stabilised telescope from external perturbations. Fine guidance is achieved with star sensors and gyros, the anticipated on axis pointing accuracy being $\sim 10''$ arc rms with additional fluctuations, proportional to the offset, due to pendulum motion about the guide stars.

Figure 1 shows the telescope assembly. To enable weaker guide stars to be used, two star sensors are to be fitted. Initially the system is locked onto a guide star of 4th magnitude or brighter using the star sensor with a 2° field of view. The second star sensor, which only has a $10'$ arc field and can operate on stars down to 7th magnitude, is now brought in by offsetting it

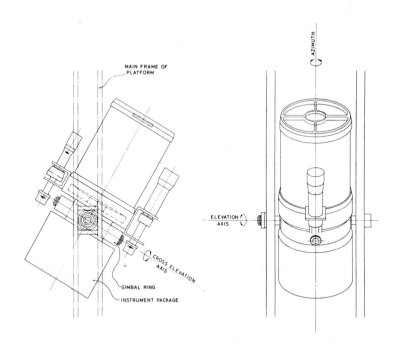

Fig. 1 Schematic to show the telescope and gimbal system for
the 60-cm telescope.

from the first star sensor and pointing it towards the second guide star.
Response at the anticipated coordinates is a good confirmation that the
correct stars are being used and the control can then be transferred from the
first to the second star sensor. The use of weaker guide stars is a great
advantage as a bright star is sometimes not available and in general it will
be possible to use smaller offset angles to the infrared object, thus
reducing the perturbations due to pendulum motion, etc.

The telescope will be used with our three band photometer system and it is
intended to include polarisation measurements for which the straight through
beam is particularly suitable. In collaboration with E.S.A. the system will
be used with a high resolution Michelson interferometer for the detection and
measurement of emission lines in the far infrared.

Beam Modulator

As on the present 39 cm telescope system, chopping on the new telescope is to

be achieved by oscillating the secondary mirror. A preliminary version of
the system to be adopted has been designed for use on the 39 cm telescope and
has incorporated some of the ideas used by Fahrbach, Haussecker and Lemke
(F.H.L., 1974) and also some of the design features employed in the telescope
system of the Groningen group.

The basic design is shown in Fig.2. The central plate has four holes for
attachment to the spider of the telescope. This plate is integral with the
aluminium columns to which are attached the 'Ling' plate and the secondary
mirror (black retaining ring) by means of flexural pivots, manufactured by
Bendix Corporation. The Ling plate has two actuators manufactured by Ling
Dynamic Systems Ltd., which pull or push via short flexible couplings so that
first the lefthand sides of the Ling plate and mirror come together, then the
right. To prevent damage to the Ling actuators due to any excessive movement

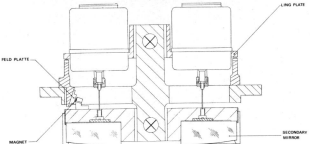

Fig. 2   Secondary mirror modulator.   Diameter $7\frac{1}{2}$ in.

which, for instance, might occur when switching on, stops are provided. These
are clear in normal operation. The diameter of the plates is $\sim$ 190 mm ($7\frac{1}{2}$ in).
In the system for the new telescope the secondary will be undersized and
supported in the centre. By doing away with the black retaining ring, etc.
which is used at present, unwanted radiation from the telescope itself will
be reduced.

The main reasoning for this design was to minimise any vibration. The natural
frequency of both the Ling plate and the mirror was made low compared to the
chopper frequency of $\sim$ 20 Hz. Many of the forces are obviously balanced and
rotational couples at frequencies well above the resonant frequencies of the
plates are almost completely 'absorbed' by the acceleration of the plates,
with the flexural pivots having only a minor effect. The final vibration
level on the telescope is extremely low. A Siemens Feld-Platte is used to
monitor the mirror movement – the magnet 'moving' over the Hall Effect plate
can be seen in Fig. 2. The output of the Feld-Platte is incorporated in a
feed-back loop which has three types of response, direct, differential and
integral as used by F.H.L. (1974). A typical chop which can be achieved is
shown in Fig. 3. This corresponds to a $6\frac{1}{2}$' arc. throw on the sky at a
frequency of 16 Hz. The squareness of the chop can be improved and operation
at higher frequencies achieved with increased power but this is undesirable

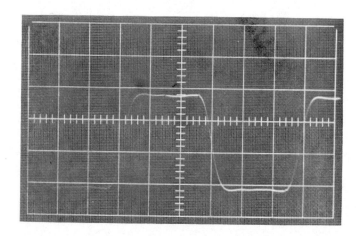

Fig. 3 Trace showing chop achieved. (Frequency 16 Hz. Throw on
sky $6\frac{1}{2}$' arc.)

in the balloon environment. The central part of the secondary which is in
its own shadow is covered by a stationary mirror, which looks back at the
detector, to reduce unwanted background radiation.

Additional features which have been incorporated in the design are:-

(1) ability to displace the centre of the chop in small steps on command from
the ground. In this way the radiation from the telescope in the two stationary positions of the secondary can be accurately balanced in flight. This is
particularly important when using a Michelson interferometer, as any
unbalanced background signals cannot be simply removed by applying a suitable
bias, as in photometric measurements.

(2) ability to hold the secondary mirror in either of the extreme positions
of the chop. The chop is obtained by switching between two voltages and
either of these voltages can be switched in continuously. This facility
would be useful to find an object and then to hold it permanently in one beam.

This beam modulator will be used on flights later this year.

Acknowledgments

The support of the design office and workshop at University College London is
greatly appreciated as is the financial support from the Science Research
Council. Mr. M. Palmer's assistance in building the electronics for the
beam modulator is gratefully acknowledged.

References

Fahrbach, U., Haussecker, K., and Lemke, D., 1974, Astron.Astrophys. 33, 265.
Tomlinson, H.S., Towlson, W.A. and Venis, T.E., 1974. Symposium Proceedings
    on Telescope Systems for Balloon-Borne Research. NASA TM X-62, 397.

DISCUSSION

Fazio What is the present status of the 60-cm telescope?
Jennings We hope to make the first flight in late summer 1976.
Schultz   What chopping frequency do you use?
Jennings   We still operate at 16 Hz, with a 6' arc throw on the sky. This is
very suitable for the balloon environment as it is well within the power
rating. However, the chopper has been run at frequencies up to 28 Hz with
a square waveform.

FAR INFRARED OBSERVATIONS WITH A 1-M BALLOON-BORNE TELESCOPE

G. G. Fazio, D. E. Kleinmann and F. J. Low

Center for Astrophysics, Cambridge, Mass. 02138, U.S.A.
and Lunar and Planetary Laboratory, University of Arizona, Tucson, Arizona 85721, U.S.A.

Abstract The Center for Astrophysics - University of Arizona balloon-borne, inertially guided telescope (102-cm) was designed to perform high resolution mapping of far-infrared (40-250 micron) sources. Three successful flights of the telescope have now occurred, resulting in 22 hours of observations on more than 40 sources. Maps with a resolution of 1' arc FWHM have been achieved with absolute position accuracies of ±10" arc. The rms noise equivalent flux density of the system is ∿70 Jy/(Hz)$^{1/2}$. From the launch site in Texas sources as far south as -50 degrees declination have been mapped.

1. Introduction

In early 1971, the Harvard College Observatory, the Smithsonian Astrophysical Observatory and the University of Arizona engaged in a cooperative programme to develop a balloon-borne, inertially stabilized, 102-cm telescope capable of carrying out far-infrared (40 - 250 μm) photometry and high resolution mapping of celestial objects. The first successful flight of the telescope occurred in February, 1974, followed by two more successful flights in May and June, 1975. A total observing time of 22 hours has now been accrued at an altitude of 29 km, and more than 40 sources have been investigated.

The telescope's instrumentation and operation have already been described in previous publications (Fazio, Kleinmann, Noyes, Wright, and Low 1974; Hazen, Coyle, and Diamond 1974; Hazen 1974). In this paper we shall briefly review the telescope properties and present the scientific results obtained to date in terms of the sensitivity and angular resolution achieved, as well as listing some of the objects that have been observed and mapped. We shall also describe several important modifications to the telescope that have markedly increased the infrared sensitivity and the efficiency of acquiring sources.

2. Instrumentation

Telescope Optics

Figure 1 shows the optical arrangement of the f/13.5 Cassegrain telescope. The 102-cm primary mirror is spherical (f/2) and constructed of an aluminum alloy; the 18-cm secondary mirror, made of pyrex, is figured to cancel the spherical aberation of the primary mirror. The Cassegrain focus occurs behind the primary mirror, yielding a scale in the focal plane of 15".2 arc/mm. At a wavelength of 100 μm, the diffraction limit of the telescope is 25" arc. Forward of the focal plane, the infrared beam is reflected by a dichroic beam splitter that passes visible light. A second beam splitter directs half the

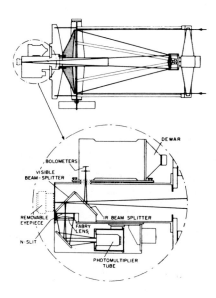

Fig. 1 Optical arrangement of the 102-cm telescope.

optical light onto an N-slit mask at a second focal plane. Light passing
through the mask is focussed onto a photomultiplier tube. A removable eyepiece, mounted in the focal plane behind the second beam splitter, serves to
aid in the optical alignment and testing of the telescope. The secondary
mirror is mounted by means of a central bolt to a solenoid-driven frame that
causes the mirror to oscillate in the azimuthal direction in a square wave
motion of 20-Hz frequency. This beam-switching technique cancels out the
background radiation from the sky and the mirror by subtracting the
contribution from two fields of view separated by 5' arc. The secondary
mirror is further mounted on a commandable focus drive.

Infrared Detection System

The infrared detection system consists of four-gallium-doped germanium
bolometers, cooled to 1.8 K in a liquid helium dewar vented to ambient
atmospheric pressure (10.5 torr at 29 km altitude). As shown in Fig.2,
three of the detectors are arranged in a linear array, with each subtending
an angle of 1!5 arc in elevation by 1' arc in azimuth; the fourth detector,
with a circular field of 0!5 arc, is located immediately adjacent to the
central detector in the linear array. The cooled optics consist of a
sandwich of crystalline quartz and calcium fluoride, with one surface coated
with diamond dust, followed by four silicon field lenses. All the cooled

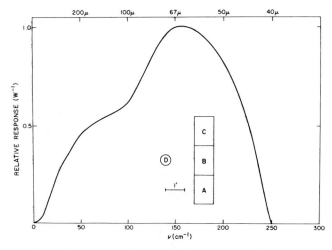

Fig. 2   Spectral response of the 1' × 1!5 arc detectors   A, B and
C.   Inset: spatial arrangement of the four detector beams.

elements are anti-reflection-coated for maximum transmission at 65 μm. The
passband of the system (Fig.2) has a sharp cuton at 40 μm, a peak
transmission of 65 μm, and a long wavelength cutoff defined primary by
diffraction. After amplification, each of the bolometer signals is digitized
and transmitted to the ground station by PCM telemetry. Phase sensitive
demodulation and further processing are done at the ground-station.

Orientation of the Telescope

The telescope is mounted in a rectangular aluminum-frame gondola (Figs. 3 and
4) 5.1 m high and 3.4 × 2.9 m wide. The entire system weighs approximately
1814 kg. The telescope is stabilized and pointed by means of positional
servo controls on the elevation and azimuth axes. The entire gondola moves
in azimuth, but the telescope motion in the elevation direction is with
respect to the gondola frame. Reaction forces for the azimuth position
control are provided by a large reaction wheel mounted on the gondola centre
line below the telescope.

Fig. 3 The 102-cm infrared telescope gondola.

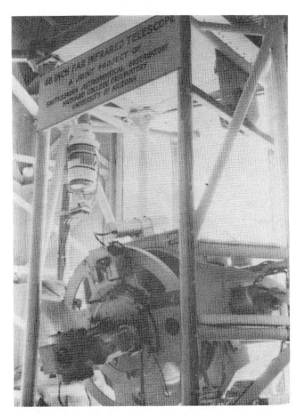

Fig. 4 Rear view of the gondola showing the photometer box at
the Cassegrain focus.

Positioning the telescope optical line of sight is accomplished in two modes:
first, an acquisition mode, determined with respect to the horizontal
component of the earth's magnetic field in azimuth and with respect to the
local vertical in elevation, to an accuracy of about $0\overset{\circ}{.}1$, and second, an
inertial mode, determined by a two-axis gyroscope system mounted on the
telescope tube. Stability in inertial space is determined by the drift rate
of the gyros, which is about a few arcmin per min of time. The gyros can be
precessed to impose angular scan rates in azimuth and elevation on the
telescope. Provisions have been made for manually commanded scan rates about
both axes as well as an onboard raster-pattern generator to produce a raster
scan of the telescope over a large field of view. The pointing control
operates most of the time in the inertial mode, and mapping is performed
using the manual and raster scans.

The N-slit photometer and a 35 mm star field camera are used for absolute position determination.

Prior to the last two flights, a television guider telescope was added to the payload for source acquisition.

3. Results

February, 1974 Flight

The first successful flight of the telescope occurred on 3 February, 1974 from Palestine, Texas. During 5.5 hours at float altitude, the telescope mapped the intensity of far-infrared radiation from the HII regions Orion A and W3 (Fazio et al., 1974 b,c; Fazio et al., 1975). In addition to Orion A the intensity and angular size of two nearby sources, the nebula M 43 and the recently discovered molecular cloud OMC-2, were also observed. For OMC-2 this data was the first observation at these wavelengths. During the mapping sequence in the region of W3, a total of six sources were observed. Only three of these sources are identified with known radio sources: W3 (G133.7 + 1.2), W3(OH), and G133.8 + 1.4. The remaining three sources were new discoveries. Of particular importance is Source 4 , a low surface brightness object that may be a large molecular cloud similar to OMC-2. During the flight, observations of Mars were used for calibration of sensitivity and angular resolution. An angular resolution of 1!25 arc FWHM was achieved for these maps and selected scans were accurate to $\sim$0!5 arc resolution. Absolute position determination was 30" arc rms. The noise equivalent flux density (NEFD) measured was $\sim$200 Jy/(Hz)$^{1/2}$. Fig. 5 compares a scan through Mars (diameter 8" arc) with a scan through W3 made made at approximately the same scan rate.

Telescope Modifications

In the February 1974 flight we not only obtained useful astronomical data, but in the process we also learned how to make the telescope more sensitive and our use of it more efficient. Through 1974, we therefore modified the telescope system to improve our control of acquisition, pointing, and operation. The telemetry commands for manual scans in the inertial mode were

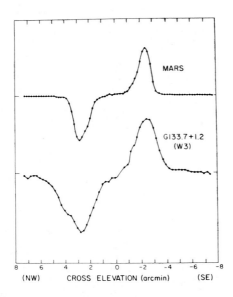

Fig. 5 Comparison of a scan through W3 along a position angle 125°
(NW to SE), with a scan through Mars (diameter 8" arc) made at
approximately the same scan rate.

simplified to one command each for up, down, left and right motion, also
giving the capability of diagonal scans. Variable scan rates were
incorporated which could be selected by telecommand. Four independent rates
(1, 2, 3 and 10' arc/sec) were available in both the azimuth and elevation
directions. The ability to reverse the direction of the raster scan mode
was incorporated. Thus, while scanning, once a source was acquired, the
raster could be immediately reversed to rescan the source. A "blip mode" of
operation was also added. In this mode, once a command was sent, the
telescope moved either in azimuth or elevation for a given length of time
and stopped. This method of operation was particularly useful in moving the
telescope in very small, but accurate, angular increments.

One of the most valuable modifications to the telescope was the addition of a
television guider telescope (Fig.6), which had a field of view of 5 degrees.
With this telescope, viewing a bright star, we were able to position the main
telescope to within 2' arc, greatly decreasing the acquisition time for an
infrared source. The television camera used was a standard closed circuit
system (Cohu Electronics, Inc., Series 2000) operating at 30 frames/sec.

Fig. 6 A view of the gondola showing the television camera mounted
in a pressurized container on the main telescope frame. The camera
is located just to the left of the aluminum cylinder containing the
balance weight.

The original vidicon was replaced with an RCA 4532 silicon diode vidicon. A
135 mm f/1.8 lens was used. The cathode circuitry of the camera was modified
to permit variable integration times on the target. The most used integration
time was about 1 sec. The video signal was used to modulate an L-band
transmitter with 1 MHz bandwidth, and at the ground station a standard 9-inch
Sony television monitor was used to display the picture. During ground tests
a star as faint as 5th magnitude could be used as the guide star, however
transmission through the telemetry degraded this performance by about one
magnitude. The television guider also permitted more rapid and accurate
inflight recalibration of the magnetometer system used for pointing in the
position mode.

A new lens and reticle for the star field camera reduced the field of view
to 5 degrees and permitted more accurate location of star positions.

We also devoted considerable effort to reducing the noise level in the
infrared signal due to improper cancellation of background radiation. A
large DC offset in the infrared signal was observed during the first flight.
All high emissivity objects, e.g. light baffles, were removed from the beam
path, the secondary mirror was undersized, a gold-coated conical mirror was
placed over the exposed bolt holding the secondary mirror, and polished brass
shields were placed over the secondary mirror support struts. The infrared
beams were centred on the optical axis to a fraction of a cm and the secondary mirror was collimated as accurately as possible.

May and June 1975 Flights

The second successful flight of the telescope occurred on 18 May 1975 and
the results of the flight showed a dramatic improvement in both the number of
objects observed and in the infrared sensitivity. During 8 hours at float
altitude some 19 sources were investigated, an increase of almost a factor of
five over the first flight. The infrared sensitivity was improved by a
factor of three. Far-infrared maps were made of the sources in the galactic
centre region, the cold, dark dust cloud near Rho Ophiucus was detected and
resolved for the first time, the H II region NGC 7538 and numerous other
regions in the southern hemisphere (e.g. RCW122, NGC 6334, NGC 6357) were
mapped, many of them for the first time.

The third successful flight occurred one month later on 16 June 1975 and was
also an extremely productive flight. During this flight some 17 source
regions were investigated, including a one-hour mapping sequence of the
galactic plane near the galactic centre. Another dark cloud source near R
Corona Austrina was found to be a bright far-infrared source; the infrared
star IRC + 10216 and the planet Uranus were detected for the first time at
these wavelengths. The galaxy M 82 was detected (Fig.7) and an attempt was
made to detect the galaxy Markarian 231. The H II regions M8, M20, M16, M17,
W31 and W33 were mapped as well as two other southern hemisphere H II regions,
G 333.6 - 0.2 and G.337.9 - 0.5. These latter two objects are at a
declination of approximately -50 degrees, and their observation exhibits that
to a large extent the southern hemisphere can be observed with a balloon-

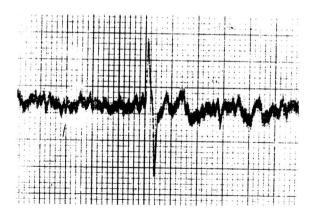

Fig. 7 Infrared detector output in real time during a scan of the
galaxy M 82. The trace is from a 1' arc detector and the
telescope was scanning at a rate of approximately 3' arc/sec in
azimuth.

borne telescope from the Texas launch site.

In summary, after modifications to the telescope, the rate of source
acquisition and mapping was increased by a factor of approximately five. The
infrared sensitivity was improved to give a NEFD of ~ 70 Jy/(Hz)$^{1/2}$. The
resolution of the telescope for the three large infrared detectors is 1.25 arc
FWHM and for the small detector 0.5 arc. The absolute source position
measurements, as determined after the flight is 10" arc rms. The ability to
integrate on a source is still limited by the DC drift rate of the azimuth
gyro, which is a few arcmin/min. At the present time integration is
achieved by repeated scans over the source.

Acknowledgments

The authors are indebted to Drs. D.E. Kleinmann, R. W. Noyes, and M. Zeilik II
for their many contributions to the design and operation of the telescope
system, and to D. Jaffe for his contribution to the television guider.
The payload was designed and constructed by the Solar Satellite Engineering
Group, Harvard College Observatory, under the direction of N. L. Hazen.
The National Scientific Balloon Facility was responsible for the launch,
tracking, and recovery of the experiment and contributed the onboard tele-
metry as well as the data-processing equipment at the ground station.
This work was supported in part by funds from the National Aeronautics and
Space Administration under grant NGR 22-007-270 to Harvard University.

References

Fazio, G.G., Kleinmann, D.C., Noyes, R.W., Wright, E.L., and Low, F.J., 1974a, in Proc.Symposium on Telescope Systems for Balloon-Borne Research, NASA Ames Research Center, Calif., NASA TM X-62, 397, p.38.

Fazio, G.G., Kleinmann, D.E., Noyes, R.W., Wright, E.L., Zeilik II, M., and Low, F.J. 1974b, Astrophys. J. 192, L23.

_____ 1974c, in Proc.8th ESLAB Symposium, ed. A.F.M. Moorwood, ESRO SP-105, p.79.

_____ 1975, A High-Resolution Map of the W3 Region at Far-Infrared Wavelengths, to be published in Astrophys.J. (Letters)., August 1975.

Hazen, N.L., Coyle, L.M., and Diamond, S.M., 1974, in Proc.Symposium on Telescope Systems for Balloon-Borne Research, NASA Ames Research Centre, Calif., NASA TM X-62, 397, p.202.

Hazen, N.L., 1974, in Proc.SPIE Seminar-in-Depth on Instrumentation in Astronomy-II, Tucson, Arizona, in press.

DISCUSSION

Jennings How did the focus of your telescope change when it got cold?

Fazio Through all our flights we have never had to change the focus from its ground based setting.

Joseph Is there a compelling reason that led you to do numerical demodulation in the ground station computer, rather than to use the usual analog method on-board or on the ground?

Fazio The digital phase sensitive detection of the signal is not done in the ground station computer, but after the flight with the computer in Cambridge. We do use the usual analog method at the ground station for real-time analysis. Phase sensitive detection was not done on board the experiment because we did not know what D.C. offset to expect.

Lemke Can you observe in the daytime?

Fazio We attempted this during one of our recent flights. The problem lies not in the infrared signal, but in the loss of all optical signals used for pointing and orientation.

Robson These are very exciting results. What flux did you quote for M 82?

Fazio I did not quote a flux, but it is in the same ball-park as that quoted by Harper, D.A. and Low, F.J. (1973, Astrophys.J. 182, L89) i.e. 1000 - 1500 Jy.

Robson Presumably you used the planets for calibration; what temperatures do you assume?

Fazio The same as everybody else in this game - those given by Armstrong, K.R., Harper, D.A., and Low, F.J. (1972, Astrophys.J. 178, L89).

Robson Finally, could you tell us what long wavelength filter you use?

Fazio We have no filter as such. We use the detector response fall-off and the diffraction of the instrument.

Moorwood In the case of the H II regions, do you find the far-infrared emission to be centred on the H II position?

Fazio Our data reduction is not yet in a state where we can answer this.

THE USE OF A LARGE TELESCOPE IN THE INFRARED

D. E. Kleinmann

Center for Astrophysics, Cambridge, Mass. 02138, U.S.A.

Abstract The impacts of the design, location, and size of a
telescope upon its infrared limitations and capabilities, and
therefore on its scientific use in the infrared are considered,
with specific application to the Large Space Telescope (LST). The
telescope, and the IR photometer proposed for it are described,
and the infrared performance to be expected from their combination
is evaluated. This performance is compared to that for other
large telescopes. Some possible far infrared astronomy applications
are indicated for the LST.

The Telescope

The LST is expected to be launched, in 1982, into an orbit with an altitude
of 610 km, an inclination of $28^{\circ}.5$, and an orbital period of 97 minutes.
It will be a 2.4 m, f/24 Cassegrain telescope having a 32% central
obscuration, and a diffraction-limited field of view of at least 5$'$ arc. The
LST will be diffraction-limited at 325 nm and longer, and it will be
maintained at 294 ± 1 K to achieve this performance. It will be slewed at
$4^{\circ}.5$/minute to within 1$''$ arc of the position commanded; a fine guidance
system can be employed on guide stars down to 14^{m} to provide pointing to
0.01$''$ arc, with an rms stability of 0.005$''$ arc.

The IR Photometer

One of the seven instruments that have been proposed for the four instrument
bays aft of the primary mirror is the infrared photometer whose design
requirements have been specified by the LST Infrared Instrument Definition
Team, consisting of Drs. G. Neugebauer, R. Hall, T. Kelsall, and the author.
Our goal was to define a sensitive, flexible photometer that would extend
the spectral response of the LST into the far infrared making it a truly

33

panchromatic telescope. The design would demand state-of-the-art detectors, filters, and cryogenic systems, assuming modest development in the state-of-the-art. Any infrared instrument would also have to contend with the relatively high background of the (warm) LST.

With these as goals, the photometer we specified will use two detector channels to cover the 2 - 1000 micron wavelength range; present plans call for either a Si:As or a Si:P photoconductor for use from 2 μm out to 24 μm or ∿ 30 μm respectively, and a bolometer for the longer wavelengths. The photometer will feature cooled interchangeable field stops and filters for each channel. The passband of the filters will nominally be 10% of the effective wavelength, although some may be as wide as 50%. The field stops will range from the diffraction limited beam size for the shortest wavelength in a channel up to 10 times the diffraction limited beam size for the longest wavelength in that channel. The detectors, filters, and apertures will be placed in a dewar and maintained at a temperature less than 2 K for an on-orbit lifetime of at least one year. Chopping will be accomplished by imparting a reciprocating rotary motion to a mirror on which the primary mirror is imaged. This chopper mirror is located in the warm relay optics forward of the dewar. The frequency of the chopper will be adjustable over 5 - 35 Hz and the amplitude will be adjustable from 0.4" arc to 210" arc on the sky.

Determinants of System Performance

Absence of the Atmosphere

For the LST there will be no atmosphere to attenuate the signal, or to produce a position-dependent, time-varying background, or to degrade the angular resolution because of "seeing". Figure 1 shows the angular resolution of several large telescopes - the LST, a 4 m telescope, the 200 inch telescope, the Multiple Mirror Telescope (MMT) operating as a phased array, and a 1 m balloon-borne or airborne telescope. A modified Rayleigh criterion has been adopted, and two points are considered resolved if their angular separation is 1/2 the diameter of the blur of light that each produces in the focal plane. The Rayleigh criterion is not physically fundamental, and two points nearer together than the Rayleigh criterion

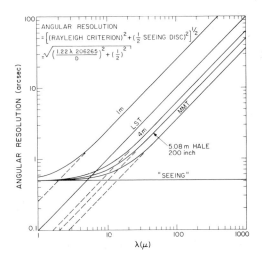

Fig. 1. The Effect of Atmospheric Seeing on Angular Resolution. Diffraction limited optics and a seeing disc 1" arc in diameter are assumed.

can be resolved if the signal-to-noise ratio is high enough. It is assumed in Fig. 1 that the contributions made to this blur by diffraction and by 1" arc seeing add as the square root of the sum of their squares. Fig. 1 shows that unless the seeing disc is smaller than 1" arc (which is considered quite good on a ground-based telescope), the LST will provide higher angular resolution than any ground-based telescope out to ∿ 5 μm. It also shows that the ground-based telescopes that are larger than the LST do not enjoy the full advantage of their size until 20 or 30 μm, where atmospheric attenuation sets in to preclude all ground-based observation out to 1 mm anyway, except for a rather occasional window at 350 μm.

Background Radiation

For an infrared instrument on the LST, the background radiation comes from the telescope itself - from the fact that the mirrors have finite emissivity. The fluctuations in this background radiation[*] are the minimum noise equival-

---

[*] See p.214f and p.288f in The Detection and Measurement of Infrared Radiation, 2nd ed., by R.A. Smith, F.E. Jones and R.P. Chasmar, Oxford Press, 1968.

ent power (NEP) for the system. This background NEP is a fundamental
limit that can be relaxed only by reducing the background on the detector,
e.g. by lowering the temperature or emissivity of the background, or by
reducing the optical throughput (AΩ), or by reducing the bandpass of the cold
filter. For the LST the temperature of the primary and secondary mirrors,
and of the telescope assembly itself, will be carefully regulated to 21 ± 1 $^\circ$C
in order to guarantee that the telescope will be diffraction limited down to
325 nm. The minimum practical throughput occurs when the system is diffraction
limited, i.e. when the beam diameter matches the Airy disc; in this case,
AΩ = $(.61 \, \pi\lambda)^2$. Note that for these diffraction limited beams, the throughput is independent of the telescope size or focal ratio, and is a function
only of wavelength.

Figure 2 shows the individual noise power of each of the most significant
contributors to the system noise power, which is the square root of the sum
of the squares of the noise power from each element. The curve LST is the
fluctuation in the background power that is incident on the detector through a
diffraction limited beam. It is the NEP that could be obtained if a perfectly
efficient detector were available. This curve was calculated for a series
of six 294 K surfaces, the first two with $\varepsilon = 0.05$ and the last four with
$\varepsilon = 0.01$. The $\varepsilon = 0.05$ surfaces are the primary and secondary mirrors of
the LST, each coated and overcoated for UV observations; the $\varepsilon = 0.01$
surfaces are mirrors in the warm part of the IR instrument, each with an
optimum IR coating. At each wavelength, the radiation incident on the
detector was assumed to have passed through a cold filter that had a 10%
bandpass and 70% transmission.

The shape of this curve should be commented upon. Where the Wien approximation is appropriate, the spectral fluctuation in the background power is the
square root of the product of the background power and the energy carried by
each photon. Where the Rayleigh–Jeans approximation is valid however, a
correction term, $(e^{hc/\lambda kT})/(e^{hc/\lambda kT} - 1)$, is introduced as a consequence of
the fact that the photons are bosons, and obey Bose–Einstein statistics. The
effect of the correction term is to increase the background NEP at longer
wavelengths; the result is that the curve is proportional to $\lambda^{-1/2}$ at longer
wavelengths rather than to λ^{-1} (which would be the case if photons obeyed
Maxwell–Boltzman statistics). Physically this difference reflects the packing

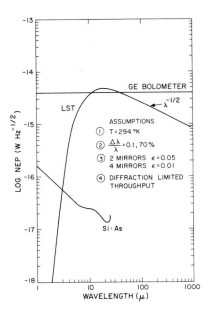

<u>Fig. 2</u>. Individual Contributions to the System Noise. Specific imputs to the calculation of the background radiation noise power conform to the properties of the LST and of the IR photometer described in the text.

of bosons into lower energy states.

Detectors

Figure 2 also shows the intrinsic noise for two detectors of possible application on the LST. There are not the only detectors of potential interest, but they are used here because they are detectors with which there is some astronomical experience. The performances indicated for these detectors in Fig. 2 are not accomplished facts; nevertheless, the projections are modest, and do not require any significant development in the state-of-the-art. The bolometer assumed in making Fig. 2 is a 2 mm diameter Ge:Ga element with a thermal conductance of 4×10^{-8} watt/^{o}K, operating at a temperature of 1.0 K (Low, private communication). This is an improvement of about a factor of three over present detectors, and it is due mostly to the lower detector temperature. The performance of photoconductors is largely determined by the background in which they operate. The noise power curve shown for Si:As shows a

sharp cutoff at 24 μm and the effect of the LST background at ∿ 10 μm.
This curve is drawn for the background expected on the LST using measurements
made under much lower backgrounds. The curve is at odds with astronomers'
experience with these detectors under higher backgrounds. The Si:As detector
has been used in filter wheel spectrometers both at KPNO and at Caltech.
The background in the spectrometer applications is about 5 - 10 times that
projected for the LST, but in these instruments the detector performance
obtained is about a factor of 30 times worse than that indicated in Fig. 2.

System Performance

The NEP is a measure of the system performance at the detector. From an
astronomical point of view, it is more useful to indicate system performance
at the entrance aperture. The noise equivalent flux density (NEFD) given in

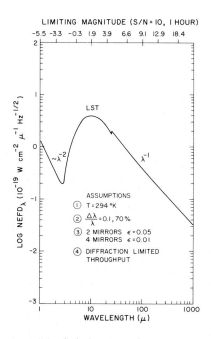

Fig. 3. NEFD and Limiting Magnitude as Functions of Wavelength.

Fig. 3 is such an indicator. The NEFD gives the flux into the telescope
that would produce an rms signal-to-noise ratio of 1 in a 1 Hz bandpass,
or equivalently, in 1/2 second of integration time. The NEFD is related
to the NEP in a straightforward way:

$$NEFD = \frac{NEP}{A \; b \; T \; D}$$

where NEP is the system NEP, obtained by taking the square root of the sum
of the squares of the noise powers from each element in the system, A is the
area of the telescope, b is the bandpass of the filter, T is the transmission of the optics, and D is a duty factor that gives the fraction of the
time that radiation from the source is actually incident on the detector. For
an ideal chopper, the duty factor would be 0.5; a duty factor of 0.4 was
used to draw the curve in Fig. 3.

The shape of the curve in Fig. 3 reflects the changing role of each of the
elements that were shown in Fig. 2. At short wavelengths out to 3 μm, the
dominant noise contributor is the Si:As detector and the NEFD is roughly
proportional to λ^{-2}. Between 3 μm and 24 μm, the fluctuations in
the background radiation dominate the system noise. The background in this
range is drawn somewhat higher in Fig. 3 than it is in Fig. 2 because the
quantum efficiency of the Si:As photodetector is expected to approach 0.5
rather than 1.0; the bolometer however, is assumed to have a quantum
efficiency of 1. The changeover between these two detectors is reflected in
the discontinuity at 24 μm. The bolometer and the background produce
comparable noise at 20 - 30 μm, but the detector dominates at longer wavelengths, where the NEFD will be proportional to λ^{-1}.

Limiting magnitudes are shown at the top of Fig. 3. These are defined to be
the visual magnitude of the A0 star (T = 9200 K) that could be measured
with an rms signal-to-noise ratio of 10 in a one hour integration. The
limiting magnitude for any specified integration time and signal-to-noise ratio
is given by

$$m_t(S/N) = -2.5 \; Log \; \left[\frac{NEFD \; (S/N)}{2.2^{\frac{1}{2}} . F^o . t^{\frac{1}{2}}}\right]$$

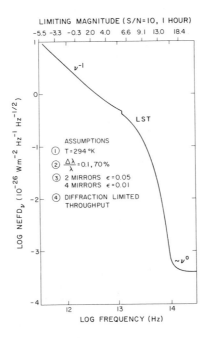

LIMITING MAGNITUDE (S/N=10, I HOUR)

<u>Fig. 4.</u> NEFD and Limiting Magnitude as Functions of Frequency.

where t is the integration time in seconds, S/N is the specified signal-to-
noise ratio, NEFD is the noise equivalent flux density at the wavelength in
question, and F^o is the flux density from a zeroth magnitude A0 star at
that wavelength. The $2^{\frac{1}{2}}$ simply converts the NEFD from an integration
time of 1/2 second (i.e. a 1 Hz bandpass) to an integration time of
one second. The factor of 2 is valid if phase sensitive detection techniques
are being used.

Figure 4 contains the same information as Fig. 3, but is drawn in the frequency
domain rather than in the wavelength domain.

<u>Comparison with other Telescopes</u>

One measure of how various telescopes compare with the LST is the integration
time they require to obtain equivalent measurements. Figure 5 provides such

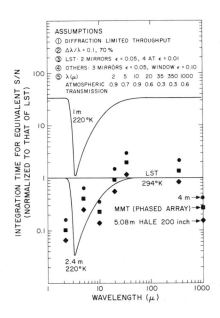

<u>Fig. 5</u>. Normalized Integration Times.

a comparison for the telescopes that were shown in Fig. 1; also shown is the integration time for a balloon-borne telescope having the same aperture as the LST. All of the integration times are normalized to those required by the LST. The same procedure used to deduce the NEFD for the LST was followed to obtain the noise equivalent flux densities of each of the other telescopes. Each of the telescopes was assumed to be equipped with a photometer using diffraction limited throughput and cold filters having 10% bandpasses and 70% transmissions. Each of the telescopes except the LST was assumed to have 3 mirrors with ε = .05 and one dewar window with ε = .10, all at ambient temperature, which was taken to be 294 K for the groundbased telescopes, and 220 K for the balloon-borne telescopes. Although the LST and to a large extent the balloon-borne telescopes can work throughout the infrared without hindrance from atmospheric absorption, groundbased telescopes can only be used in atmospheric "windows" where there is measurable transmission. The normalized integration times for the groundbased telescopes are therefore shown as discrete points in Fig. 5. The atmospheric windows, and the transmission through them at a dry site are tabulated in the legend in Fig. 5.

Figure 5 raises a number of very interesting points. First, it emphasizes
the limited infrared spectral range that is available from the ground. Second,
it demonstrates that the LST will in some instances require less integration
time than much larger ground-based telescopes, both because seeing often
precludes the use of diffraction limited beams on ground-based telescopes,
and because the transmission in some of the atmospheric windows is a poor
and sometime-thing at best that is variable from night to night, and even
within a "photometric" night. Third, the figure dramatically indicates the
benefits to be gained from reasonably small decreases in background temperature. Fourth, by the same token, it shows that for the temperature, filter
bandpasses, and throughput involved in the LST, the performance short of
3 μm and longward of ∿ 30 μm is already detector limited. This means
that in these two regions somewhat larger throughputs and/or filter bandpasses
can be used without seriously affecting the sensitivity. Finally, the 91 cm
airborne telescope in NASA's C-141 may at some time approach the performance
indicated for a 1 m telescope in Fig. 5, but because of the pointing
requirement it seems unlikely that 1 m and 2.4 m balloon borne telescopes
ever will, except perhaps beyond ∿ 50 μm. Even at this wavelength, the
pointing required to make long integrations with a diffraction limited beam
is difficult to achieve on a balloon platform. The flight-ready weight for
a 2.4 m balloon telescope stabilized to 1" arc could easily exceed the
5000 kg that can be launched at the present time with high reliability to
29 km.

Some Possible Application in Far Infrared Astronomy

Figure 6 puts the performance of the LST into astronomical perspective by
addressing the question, "How far could some of the bright, well-known
far-infrared sources be removed before the LST instrument would require 1
hour to make a 10 sigma measurement?" Galactic objects are shown on
the light in Fig. 6 and extragalactic objects are shown on the right. Several
distance milestones are indicated by the dashed horizontal lines, each of
which is labelled on the right. The actual distance of each source is
indicated by the X on its line. The actual distances of NGC 2024 and
Orion, .0004 Mpc and .0005 Mpc respectively, could not be accommodated
within the boundaries of this figure. Far infrared flux measurements (preferably those with effective wavelengths between 65 μm and 100 μm) were

compiled from the literature (Hoffmann et al 1971, Harper and Low 1971, Harper and Low 1973, Emerson et al 1973, Fazio et al 1974, Olthoff 1975). Note however that neither 3C 273 or Markarian 231 have been measured at ∿ 100 μm. An estimate of the 100 μm flux from 3C 273 was obtained by interpolating between the ∿ 20 Jy seen at 3 mm (Fogarty et al 1971) and the 0.4 Jy seen at 10 μm (Kleinmann and Low 1970). The far infrared flux from Markarian 231 was estimated by noting that its 10 - 20 μm spectrum strongly resembles that of M 82 (Rieke, private communication), and by assuming that the resemblance continues out to 100 μm. The ratio of their 10 μm fluxes is about 20 (Rieke and Low 1972), so the expected 100 μm flux for Markarian 231 would be about 1/20 as bright as that observed for M 82. The list of fluxes so compiled was compared to the NEFD drawn in Figs. 3 and 4 to determine the signal-to-noise ratios that would be obtained on these objects by the LST in a 1/2 second integration. Then, the distance, d, at which a specified signal-to-noise ratio, S/N, would be obtained in a specified time, t (in seconds), is given by

$$(\frac{d}{d_o})^2 = \frac{\frac{\text{Flux}.2.2^{\frac{1}{2}}}{\text{NEFD}}}{\text{S/N}} t^{\frac{1}{2}} .$$

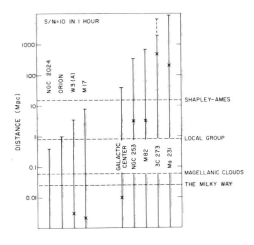

Fig. 6. The Far Infrared Observability of Galactic and Extragalactic Sources.

Figure 6 suggests several intriguing observational programs. It would be possible for example to determine the number and location of all the sources similar to the Orion nebula, W3(A), or M 17 in each of the galaxies in the local group. It should be possible to monitor the quasi-stellar source, 3C 273, to determine whether there is variability at 100 µm, and if so, how it compares with that reported at 3 mm. It would be possible to make a far-infrared survey of all 1149 galaxies in the Shapley-Ames catalogue to determine which ones have phenomena like that found in the Galactic Centre.

Acknowledgements

It is a pleasure to thank Drs. S. G. Kleinmann, G. Neugebauer and G. G. Fazio for reading and commenting upon a preliminary manuscript. A discussion about detectors with Dr. F. J. Low was very helpful. Much of the computational work was performed by Mr. J. L. Geary.

References

Emerson, J. P., Jennings, R. E. and Moorwood, A. F. M., 1973, Astrophys.J., 184, 401–414.
Fazio, G. G., Kleinmann, D. E., Noyes, R. W., Wright, E. L., Zeilik, M. II. and Low, F. J., 1974, Proc. Eight ESLAB Symposium, Frascati, Italy, pp.79–85.
Fogarty, W. G., Epstein, E. E., Montgomery, J. W., Dworetsky, M. M., 1971, Astron.J., 76, 537–543.
Harper, D. A. and Low, F. J., 1971, Astrophys.J., 165, L9–L13.
Harper, D. A. and Low, F. J., 1973, Astrophys.J., 182, L89–L93.
Hoffmann, W. F., Frederick, C. L. and Emery, R. J., 1971, Astrophys.J., 170, L89–97.
Kleinmann, D. E. and Low, F. J., 1970, Astrophys.J., 161, L203–L206.
Olthoff, Jenk, 1975, Ph.D. dissertation.
Rieke, G. H. and Low, F. J., 1972, Astrophys.J., 176, L95–L100.
Smith, R. A., Jones, F. E. and Chasmar, R. P., 1968, The Detection and Measurement of Infra-red Radiation, 2nd edition, Oxford Univ. Press.

DISCUSSION

Marsden How much liquid helium is to be carried for infrared experiments on the LST?
Kleinmann The Ball Brothers design requires 13 kg of He and 50 kg of SN_2. The Lockheed design requires 25 kg of He.
Shivanandan What will the cryogenic endurance be on the LST for infrared detectors?
Kleinmann At present we expect a lifetime in orbit in excess of one year. Both competing industry teams (Ball Bros. and Lockheed) and the independent consultants at Arthur D. Little conclude that a one year lifetime is feasible at the weight and price limits of the infrared equipment on the LST.
Shivanandan What is the uniqueness of the LST for a 100 µm sky survey,

as compared with the Fazio and Jennings balloon-borne telescopes.

Kleinmann The uniqueness of the LST lies in the angular resolution and in the sensitivity which can be achieved because of its large size. The LST will be useful for studies of individual objects rather than for a sky survey.

van Duinen When you were talking about diffraction limited resolution, were you referring to the size defined by the Airy disc?

Kleinmann Yes.

van Duinen Did you take into account the distortion caused by the central obscuration.

Kleinmann No.

van Duinen Your NEP (noise equivalent power) figures took into account emission by the mirrors only. Is effective suppression of other sources of emission like the black central hole possible?

Kleinmann Yes.

Melchiorri The telescope has been studied from an optical point of view. In the millimetre region radio effects, such as Fresnel diffraction, may be important. Have you computed the efficiency of the telescope on this basis?

Kleinmann No.

Coron What sort of sky chopping, and at what frequency, do you plan on the LST infrared system?

Kleinmann We have specified a focal plane chopper with a throw adjustable from 0.4" arc to 210" arc at a frequency continuously adjustable from 5 to 35 Hz. The stability of the two end positions will be 0.04" arc or 1% of the throw, whichever is the larger; we would prefer a stability of 0.04" arc independent of the beam separation.

Jelley Will the LST be used in 'daytime' and, if so, will the effective background be any higher?

Kleinmann The LST will not be used within 25° of the sun. Outside this zone of avoidance the LST will be used in 'daytime'. Because of the sun-shading and the thermal control on the LST, the effective background will be about the same in 'daytime' and 'nighttime'.

Jelley Will the stellar magnitude limit on the guide star be any different in 'daytime' work, and do you guide on the visible or infrared part of the guide star spectrum?

Kleinmann We will guide on visible stars, and the magnitude limit for these will be the same day or night.

Fazio What will the limiting flux at 100 μm be for the LST and how does this compare with what can be done from balloons?

Kleinmann At 100 μm the NEFD (noise equivalent flux density) for the LST will be about 1 Jy Hz$^{-1/2}$. Since you are operating very near the background limit, the NEFD for balloon systems will scale approximately as the area of the primary mirror, provided that the pointing of the balloon telescope is good enough to use the diffraction limited beam.

Joseph If the infrared system survives the LST payload redefinition, how do you see the participation of guest observers taking shape?

Kleinmann Philosophically, I personally look for (i) equitable distribution of time to the Principle Investigator and his co-investigators, (ii) early inclusion of outside guest investigators.

CHOPPING PRIMARY FOR A BALLOON-BORNE IR TELESCOPE

D. Lemke and K. Haussecker

Max-Planck-Institut fur Astronomie, D-69 Heidelberg, W. Germany

Abstract The design of a servo controlled chopper driver for a 20-cm primary of a far infrared telescope is described. The system allows square wave modulation even at large chopper throws without causing vibrations or acoustic noise. The reliability and temperature stability of the system have been tested during a recent balloon flight.

We have constructed a small balloon-borne far infrared telescope around a basic idea of Dr. F. J. Low. The concept is extremely simple: the telescope is made of only one part, a chopping 20-cm primary. The detector is placed off-axis near the f/4 focus. The purpose of this instrument is to survey the sky in the far infrared (50 μm < λ < 250 μm) with a very large beam (40" arc).

It is not easy to wobble the massive 20-cm mirror at a frequency in the 10 - 20 Hz region, especially if one intends to achieve:

- a modulation function close to a square wave,
- stable end positions and throw, independent of the attitude of the telescope,
- no vibrations which may introduce noise due to the microphonics of the detector or the preamp,
- and finally a complete electronic remote control of the chopper throw and frequency.

We have found a satisfying solution for this task by introducing the concept of servo control into the chopper design. The main parts of this arrangement can be seen in Fig. 2. The mirror is glued to steel hinge spring clamps. One of these springs is mounted to a second spring crossed by 90° to allow for different thermal expansion of the mirror and the

<u>Fig. 1</u> The 20-cm infrared telescope mounted on the balloon-borne telescope THISBE.

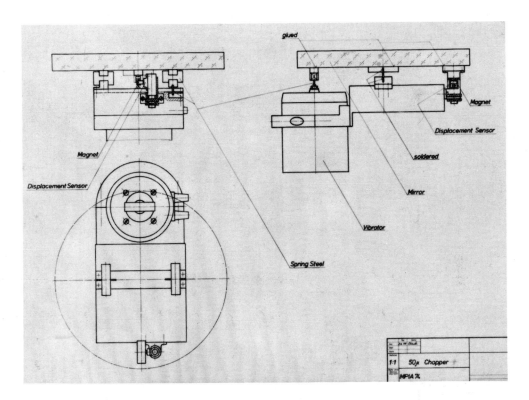

Fig. 2 Mounting of the chopping primary.

mounting. The mirror is driven by a Ling Vibrator Model 200. The movement
of the chopper is measured by a displacement sensor, a Siemens Feldplatte
FP 200 L 100.

Figure 3 shows the simplified circuits: the difference signal between the
square wave produced by the signal generator G and the actual modulation
signal measured by the Feldplatte DS is fed into three parallel circuits
P, D, I. Proportional amplification P and differentiation D control the
shape and fast rise of the modulation function. Integration I controls the
zero position and consequently the constancy of the end positions of the
chopper.

Figure 4 shows the modulation function of the 20 cm primary with throws of
50' arc and 5' arc. With the servo control "off" the modulation is some sort
of a "sinus". With the servo "on" the modulation function is close to a

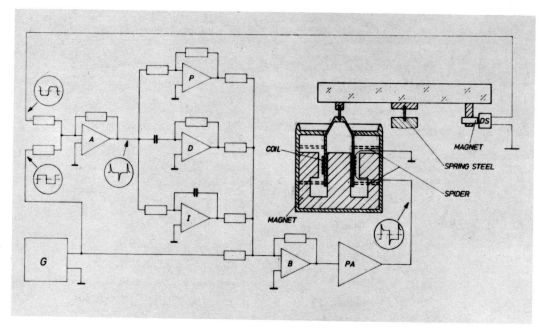

<u>Fig. 3</u> Simplified circuit of the servo controlled chopper. DS
displacement sensor, G square wave generator, PA power
amplifier.

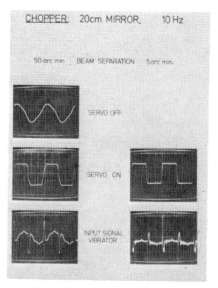

<u>Fig. 4</u> Modulation functions of the 20-cm chopping primary for a
very large and a moderate throw.

square wave with the shape depending somewhat on the beam separation.
Finally, Figure 4 bottom shows the vibrator input signals: they have a
very fast rise to accelerate the mirror fast, but before the mirror reaches
one of its end positions, the signal becomes negative to stop the mirror
just at this point without flutter.

The chopper works smoothly and quietly because there are no mechanical
buffer stops defining the end positions. No acoustic noise was measured in
the detector circuit. All parameters of the modulation (shape, frequency,
beam separation) can be changed quickly by tuning the electronics. With
selected displacement sensors the system works almost independently of the
temperature over a wide range allowing balloon-borne and ground-based
applications. All parts of this chopper arrangement are cheap and
commercially available.

The chopping primary was tested for the first time on a balloon flight in
autumn 1974. The main parameters were monitored and controlled from the
ground. No deviations from the laboratory results were found.

The arrangement described here seems not only to be useful for balloon-borne
experiments, but also for the larger and heavier chopping secondaries of
large ground based telescopes. We will use it on several telescopes of the
MPIA on Calar Alto.

DISCUSSION

Furniss What is the power consumption for the 50' arc chop?
Lemke About 15 W for the chopper and the servo electronics with 4 W fed
into the vibrator. The total consumption may be reduced by a factor of 2
or so with a more carefully designed power amplifier.
A. Harris Could you describe your mirror mounting in more detail?
Lemke The mirror is glued on its back by a special mixture of epoxy
flexible at low temperatures to two blocks of Invar. Invar has
approximately the same thermal expansion as the Duran mirror. The Invar
blocks are soldered to steel hinge spring clamps. One of these supports is
mounted to a second steel spring crossed by 90° allowing different thermal
expansion of the mirror and the metallic support. No precautions have been
taken to arrange the Ling vibrator completely behind the primary because
the mirror is surrounded by a large radiation shield.

FOURIER-SPECTROSCOPY FROM BALLOON PLATFORMS

R. Hofmann, K. W. Michel, F. Naumann and J. Stocker

Max-Planck-Institut für Extraterrest. Physik, 8046 Garching München, W. Germany

Abstract The requirements and design of a balloon-borne Fourier
spectrometer are described in view of limited pointing accuracy
and suppression of atmospheric line radiation by the wobbling
mirror technique. Because of the low efficiency of a Michelson
over a larger wavelength region (20 - 200 μm), a lamellar grating
for stepwise operation has been constructed, the mechanical design
of which meets the accuracy requirements (< 1 μm) over a stroke
length of 10 cm. A preamplifier with variable time constant
provides the dynamics necessary to allow a reasonable duty factor,
even if compensating atmospheric radiation exceeding interstellar
line fluxes by a factor of 10^3. The effect of finite pointing
accuracy is treated quantitatively.

1. Specifications of the Experiment

We wish to study interstellar objects by Fourier spectroscopy in the spectral
range from 20 to 200 μm which later on can be extended to the region of
about 500 μm (1). The spectral resolution is 0.05 cm^{-1} which means that
we need an interferometer with 10 cm stroke length. The step width of the
interferometer is determined by the shortest wavelength and will be 5 μm for
the first experiments. So the number of sampling points in the interferogram
and hence the number of spectral elements is of the order 10^4. Using a
stepping frequency of 10 Hz, the observation time for one interferogram is
about 30 min. For this time, the weakest detectable radiance per spectral
line is 4.10^{-11} W cm^{-2}sr^{-1}, if the diameter of the radiation source is
greater than or equal to the field of view of the telescope. (Diameter of
the primary mirror: 1 m, field of view: 1' arc, NEP = 3.10^{-14}W/Hz.

A block diagram of the data acquisition system for the experiment from the
first modulation unit to the point where digital data are available is shown
in Fig. 1.

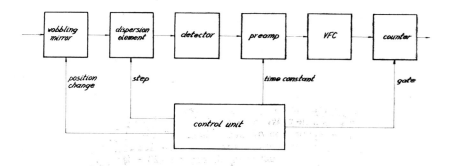

<u>Fig. 1.</u> Block diagram of the data acquisition system.

In the following, problems concerning the dispersion element, the preamplifier
are described together with the effect of finite pointing accuracy on the
spectrum obtained from the measured interferogram.

2.   The Lamellar Grating

The dispersion element is a lamellar grating interferometer (2) instead of the
more common Michelson interferometer, because of the higher efficiency over
the whole spectral range. The efficiency of the Michelson interferometer
is mainly determined by the optical constants of the beam splitter material,
while that of the lamellar grating is completely determined by diffraction
and path difference effects (3). A measure for the efficiency is the degree
of modulation, which is shown in Fig. 2 for a lamellar grating for several
grating constants α and beam divergences 2ψ at the grating as a function
of the wavelength. F_{min} and F_{max} are the fluxes at the detector for
destructive and constructive interference. The calculated performance of
our instrument (α = 0.5 cm, 2ψ = 10' arc) is represented by curve 1, which
shows a cut-off at about 15 μm. In order to obtain this performance, the
maximum path difference of interfering beams has to be small as compared to
the smallest wavelength used, that is of the order 2 μm. Therefore the
guiding system, which must work under balloon conditions, is rather complicated,
as can be seen in Fig. 3. The 20 movable lamellas, made as well as the
fixed ones from ZERODUR, a ceramic with extremely low thermal expansion,
are gliding on two optically flat ledges made of the same material with four

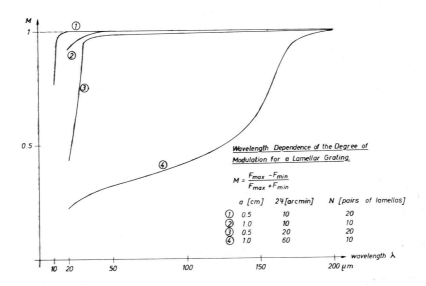

Fig. 2. Wavelength dependence of the degree of modulation for a
Lamellar grating.

fine adjustable pads, driven by a stepper motor via a high precision pendulum
supported micrometer screw. Position measurement is done by an inductive
scan system with an accuracy of ±0.5 m. With this system, the inclination
of the plane of the movable lamellas relative to that of the fixed ones is
smaller than one arc sec for any path difference.

The instrument is constructed for operation at liquid nitrogen temperature.
Its main disadvantage is that it consists of forty lamellas, each 2.5 mm
thick, which have to be adjusted with high precision, while the Michelson
interferometer has only three parts (two mirrors, one beam splitter). An
optical flatness over each package of 20 lamellas of better than a wavelength in the visible has been achieved.

3. The Preamplifier

One big problem in Fourier spectroscopy is the suppression of the background

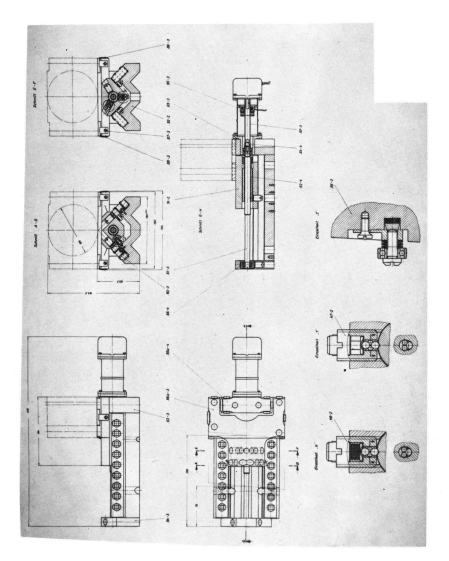

Fig. 3. The interferometer guiding system.

radiation. The ratio of the radiance of the weakest detectable line to
that of the background radiation in the same spectral element is of the
order 10^{-3}. Therefore three orders of magnitude of the background flux
have to be suppressed in order to see this line. The suppression is done
by sky chopping with the wobbling mirror.

The efficiency of this operation is determined by several parameters, the
most crucial of which is the settling time constant τ of the AC preamplifier. It determines how the out-put signal falls off for a constant input
signal. Figure 4 shows the consequence of this behaviour for our experiment.
Assume that the telescope is looking at the background and the interferometer
makes one step which causes a flux change at the detector and a signal change
at the preamplifier output, which is completely due to the variation of the
background interferogram (Fig. 5). This signal falls off exponentially
with the time constant τ. Now one has to wait until this signal
change becomes small compared to the signal of the interstellar IR-source,
before one can move the wobbling mirror and make a measurement. Since the
ratio of the background flux to the source flux can be very large, much
time is lost for this settling process if τ is large. If τ is small
the signal of the infrared source falls off too steeply during integration.
In both cases the efficiency is low. Therefore, we decided to use a
preamplifier with two changeable time constants. The small one, τ_1, is used
during the step of the interferometer, until the signal has fallen off to
nearly zero level. Then the large one, τ_2, is used during the signal
integration and the motion of the wobbling mirror. Figure 6 shows a scheme
of our solution for the preamplifier. The time constants are changed by
switch S_3. The small one is of the order of the bolometer time constant,
that is some milliseconds, the large one is some hundred milliseconds.

4. Effects of Finite Pointing Accuracy

One basic assumption in Fourier spectroscopy is that the flux from the
radiation source is constant during the integration time of the interferogram.
But we want to look at extended sources and the balloon-borne telescope has
a finite pointing accuracy. The effect is, that the measured signal is
modulated due to flux changes generated by the motion of the telescope

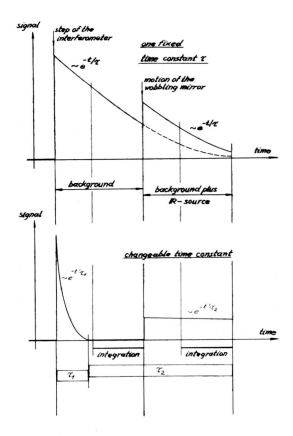

<u>Fig. 4</u>. The effect of the preamplifier time-constant on the output signal.

axis. The intensity of these satellite bands and the intensity loss of the mainline depend on the mean amplitude of the pointing-error, which thus should be kept appreciably smaller than the field of view of $1'$ arc. This illustrates the high requirements on the attitude control system in balloon-borne Fourier spectroscopy (4).

What happens can best be understood if this motion is assumed to be harmonic and has the frequency f_p. The source spectrum consists of one narrow line with flux B_o at the wave-number $\tilde{\nu}_o$: $B(\tilde{\nu}) = B_o \cdot \delta(\nu - \nu_o)$. The flux

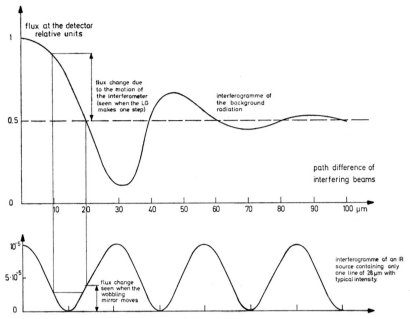

Fig. 5. The flux change at the detector due to the interferometer motion.

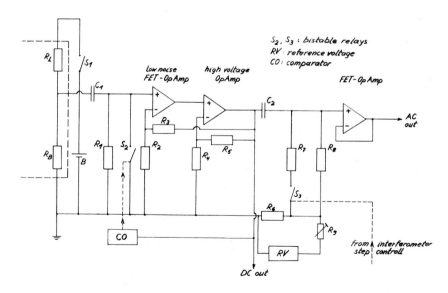

Fig. 6. Proposed preamplifier scheme.

of this line at the detector is modulated by the telescope-oscillation with the relative amplitude ε, which gives the signal modulation:

$$s \sim B_o \cdot \delta(\tilde{\nu} - \tilde{\nu}_o)(1 + \varepsilon \cdot \cos(2\pi f_p t + \rho)). \tag{1}$$

For simplification, assume that $\rho = 0$. t can be substituted by $x/2v$, where x is the path difference of interfering beams after the dispersion element and v is the mean stepping velocity of the dispersion element, then (1) becomes:

$$s \sim B_o \cdot \delta(\tilde{\nu} - \tilde{\nu}_o)(1 + \varepsilon \cdot \cos \frac{2\pi f_p x}{2v}) \tag{2}$$

And the Fourier transform is

$$F(x) \sim B_o (1 + \varepsilon \cdot \cos \frac{2\pi f_p x}{2v}) \cos 2\pi \tilde{\nu}_o x. \tag{3}$$

The inverse transform gives a spectrum which now consists of a central at wavenumber $\tilde{\nu}_o$ with amplitude B_o and two satellite lines at wavenumber $\tilde{\nu} = \tilde{\nu}_o \pm f_p/2v$ with amplitude $\frac{\varepsilon}{2} B_o$.

For example, in our experiment $2v = 10^{-2}$ cm s^{-1}. Let $f_p = 1$Hz, which is the order of magnitude of the resonance frequency for our pointing system, then the wavenumber shift of the satellite lines relative to the central line is $f_p/2v = 1$ cm^{-1}, which is large compared with the spectral resolution of 0.05 cm^{-1}.

Usually the telescope motion will show a certain frequency spectrum. In this case, the satellite lines are washed out to become sidebands which are exact images of this spectrum. In the limiting case of statistical motion, a new type of noise is introduced in the spectrum.

6.   Conclusions

The problems just described lead to the following conclusion:

i)   For taking IR-spectra over a wide spectral range, the lamellar grating
      seems to be the most effective interferometer.

ii)   High resolution spectroscopy with high time economy as necessary for
      balloon-borne instruments requires stepwise operation of the inter-
      ferometer and a preamplifier with changeable time constants for
      suppression of the background emission.

iii) Fourier spectroscopy requires high pointing accuracy if effects of
      the telescope motion on the measured spectrum should be avoided.

References

(1)   F. Naumann, K. W. Michel, 1973, Hochauflösende IR-Spektralphotometrie
         mit einem Ballon-teleskop, MPI - PAE/TB Extraterr. 9.
(2)   J. Strong, G. A. Vanasse, J.Opt.Soc.Am., 50, 113, 1960.
(3)   R. Hofmann, 1974, Das optische System eines Ballonteleskops für
         hochauflösende Fourier Spektroskopie im infraroten Spektralbereich,
         MPI - PAE/Extraterr. 103.
(4)   L. Haser, 1974, Konzept der Lageregelung für ein Infrarot-Ballonteleskop,
         MPI - PAE/TB Extraterr. 12.

DISCUSSION

Fazio What is the status of the Max Planck Institute's program of balloon-
borne far infrared astronomy?
Hofmann I think the telescope will be assembled at the end of 1976. Then
the first technical projects can be done.

THE MOON AS A CALIBRATION SOURCE FOR SUBMILLIMETRE ASTRONOMY

M. J. Pugh

Queen Mary College, London E.1

Abstract   A limited solution to the problem of thermal emission
in the presence of simple scattering is described. This is used
with the temperature profiles from Apollo 17 to produce brightness
temperature curves along the thermal equator. These are proposed
as standards to overcome the difficulties of absolute measurements
through the terrestrial atmosphere.

Given the high levels and complexity of atmospheric absorption at sub-
millimetre wavelengths, absolute measurements of brightness temperatures
become almost impossibly difficult. A new approach to the solar brightness
is described elsewhere in these proceedings by B. Carli, but the present
paper is concerned with a theoretical model of the lunar brightness.

The moon has been proposed many times as a calibration source, especially in
recent times by Linsky (1973), but the latter proposal only covers the mean
temperature over a whole lunar cycle. In terms of observational measurements
this is rather difficult to use; continuing observation spanning several
cycles is needed to calibrate some easier object (e.g. a planet) as a
transfer standard. Variation of water vapour content in the atmosphere at
any given site has a drastic effect on the levels and wavelengths of
observation. This effect is illustrated in Table 1, taken from unpublished
calculations by the author. The models of the filters used are taken from
P.Ade, and applied to calculated transmission spectra of a 150 K thermal
source, seen through a model atmosphere for Izana, Tenerife, where the
average water content is 2 mm, although good and rare excellent days can
realise 1 mm and 0.5 mm respectively. It will be seen that signal levels can
fluctuate considerably, and reduction is complicated by the movement of mean
frequency. No further consideration of these calculations is contemplated
here, but they do illustrate the great desirability of "instantaneous"
calibrations for brightness temperatures, which may even be used to monitor

atmospheric absorption fluctuations.

Clegg et al. (1972) demonstrated the existence of a density-dependent
component of the attenuation coefficient per unit mass of lunar fines. This
is tentatively ascribed to scattering of radiation within the sample. Further
work on lunar samples by the author has confirmed the existence of a
scattering component, and assigned functional forms to the variation of the
true absorption and scattering attenuation components, with respect to the
variables of temperature, frequency and density within the submillimetre
wavelength range. With the availability of the Apollo 17 thermal profile
down to 0.66 m, tabulated at 6° intervals in phase, provided by personal
communication with S. J. Keihm, it becomes possible to construct the
observable brightness temperatures at the phases of tabulation. This
construction must take into account the existence of scattering when setting
up the source function to be entered in the equation of radiative transfer,
as well as including its effect in the attenuation. Equation (1) states
the solution to the equation of radiative transfer, and may be taken to
define the quantity $\maltese$, both in its dimensions and in the way it is to enter
into the equation. $\maltese$ is the contribution from scattering to the source
function. The other quantities used are B, the Planck source function, ρ,
the density κ, the absorption coefficient, α_D, the scattering attenuation
coefficient (the last two are per unit mass). Distance z is measured from
z = 0 at the lunar surface, with positive direction downwards.

$$I_{z=0} = \int_0^z \rho(\kappa B + \alpha_D \maltese) \exp(-\tau_{z,0}) \, dz + I_z \exp(-\tau_{z,0}) \qquad (1)$$

$$\text{where} \quad \tau_{a,b} = \left| \int_b^a \rho(\kappa + \alpha_D) \, dz \right|$$

The source function contribution $\maltese$ is given in Eq. (2) for the scheme of
calculation used here. This represents simply the integral over all space of
the thermal emission from the material, suitably attenuated by the distance
between sites of emission and scattering, and singly scattered into the
normal emergent beam (this beam proceeds along the (-z) -axis). It must be
emphasised that here is the first major approximation made; the scattering
is considered as <u>single</u>. Also introduced here is the scattering cross-
section $\sigma_D(\phi)$.

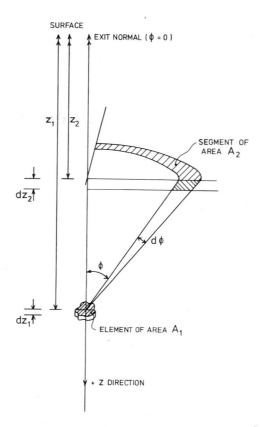

<u>Fig. 1</u>. Geometry used to derive Eq.(2). The energy received by
area A_1 from all areas A_2 is calculated, and converted to the
intensity $\mp$ scattered along $\phi = 0$.

$$\mp (z_1) \;\; = \;\; \int_0^z \rho_2 \; \kappa_2 B_2 dz_2 \int_0^{\pi/2} \frac{2\pi\sigma_D(\phi)}{\alpha_D} \;\; \exp(-\tau_{z_1,z_2} . \sec \; \phi) \;.\sin \; \phi.d \; \phi$$

where

$$\alpha_D \;\; = \;\; \int_{4\pi} \sigma_D(\phi).d\Omega.$$

The geometry of this integral is indicated in Fig. 1 from which it may be
derived quite simply. It will be noted that another assumption has been made,
that the scattering cross-section depends on only one angular coordinate.
The other, which covers the orientation of the particles, is assumed to
average out over the whole assembly of particles and their orientations
within the lunar surface layer.

Equations (1) and (2) may be solved by inserting all the relevant functions
and proceeding as appropriate. In certain cases, suitably simplified, this
can be done analytically with the aid of tables of integrals. This the
author has done for simple forms of $\sigma_D(\phi)$ with ρ, κ, α_D all constant with
depth and B either constant or linear.   In Eq.(2) it becomes necessary to
effect a change of operator:

$$\int_0^z dz_2 \quad becomes \quad \int_0^{z_1} dz_2 \quad + \quad \int_{z_1}^z dz_2$$

and for analytical solutions the limit $z \to \infty$ is taken as a convenient
simplification to dispose of the second term in Eq.(1). For these simple
cases, and taking the extreme case $\alpha_D = \kappa$ (possible at low densities
according to Clegg et al. 1972), we find the ratio $\mp_{z=0}/B_{z=0}$ is between
61% and 63% plus approximately 64% of the thermal gradient per unit optical
depth.

For more complicated functions such as those derived by the author from
sample measurements where α_D depends on ρ and T, κ depends on T, ρ
and T depend on z, the analytic solution becomes daunting. Taking ρ
from Carrier et al. (1973) and using power law approximations to interpolate
temperature T in the data of Keihm, the problem was rapidly converted into
a simple Algol-60 program to run on the University of London Computer Centre
CDC6600 machine. The integrals were accomplished using the N.A.G.
algorithm D01AAA. Algol is necessitated by the fact that the solution is
expressible as a triple integral form, involving recursive use of the
integral segment.

This machine solution took up about $14\frac{1}{2}$ K words of core, and used some 45
seconds to compute each output number. Brightness temperatures were
computed at six frequencies (frequency (in cm^{-1}):= 5 step 5 until 30), but
only three are shown in Fig.2, as all the curves lie conformally in order,
and clarity is required in the figure. The dashed curve is the surface
temperature. The observable brightness temperatures are not simply constant
ratios of the surface temperature; the emissivity is some $1\frac{1}{2}$% to 2% higher
during the lunar day than during the night, at all frequencies. The effect
of the varying sign of the immediate subsurface temperature is much less
pronounced than originally expected.

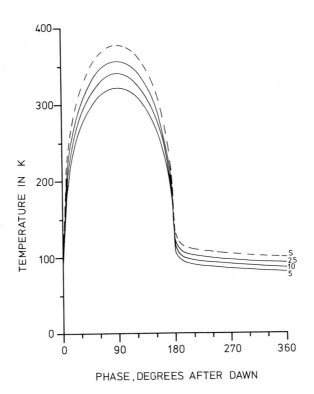

Fig. 2   Calculated brightness temperatures of the Moon at frequencies of 5, 10 and 25 cm^{-1}. Also shown is the surface temperature (dashed curve).

TABLE 1

Source: Thermal object at 150 K

| Water content | 3 mm. Black Polythene | | "350 µm" Interference Filter | |
	ENERGY	FREQUENCY	ENERGY	FREQUENCY
10	26	8.0	0	22.4
5	40	9.6	6	26.2
2	66	13.4	60	26.5
1	115	17.1	200	26.9
0.5	176	19.6	416	28.2

Energy in 10^{-9} W.m.$^{-2}$ster.$^{-1}$ Frequency in cm.$^{-1}$

In conclusion, it is proposed that this calculation, although simple in
scope and treatment, and not susceptible to generalisation, forms the basis
for using the moon as a calibration source for far-infrared astronomy, by
taking equatorial lunar scans for calibration.   The curves of Fig. 2 need
to be operated on to produce expectation curves for various beam sizes of
telescopes, and for all reasonable lunar phases, e.g. intervals of about two
days.

The suitability of the moon should be beyond question, since the surface
properties, such as density and temperature profiles have been measured in
situ, and the properties of the material can be measured, and remeasured, in
terrestrial laboratories.   There is, as yet, no other astronomical body for
which that is true. Also, the moon is usually available when far infrared
astronomers have time on large telescopes, as the light level when the moon
is above the horizon is unsuitable for optical astronomers, who then gladly
relinquish the telescopes to others.

Disadvantages would arise for such telescopes as have antenna patterns large
in comparison with the lunar size; a model of the poleward reduction of
temperature is required, and then convolution of the lunar brightnesses with
the telescope pattern is required to provide synthetic equatorial scans.

References

Ade, P.A.R., 1973, Ph.D. Thesis, University of London.
Carrier, W.D. III, Mitchell, J.K., Mahmood, A., 1973, Proc. of the Fourth
     Lunar Science Conference, Supp.4, Geochimica et Cosmochimica Acta 3,
     2403-2411.
Clegg, P.E., Pandya, S.J., Foster, S.A., Bastin, J.A., 1972, Proc. of the
     Third Lunar Science Conference, Supp.3, Geochimica et Cosmochimica
     Acta, 3, 3035-3045.
Linsky, J.L., 1973, Astrophys.Supp.Ser. No.216, 25, 163-204.

DISCUSSION

Kleinmann Besides lunar phase, I suspect that selenographic position may be
very important in using the moon either as a short term intermediate reference
- for atmospheric extinction, variability - or as an absolute calibrator.
    This information is vital at very much shorter infrared wavelengths
(work by UCSD group - Soifer, Russell et al., recent Astrophys.J.), where the
flux varies measurably with selenographic position. I do not know what the
spectral behaviour of the lunar surface is, but the point should be

investigated.

Pugh There do indeed exist features visible on the moon at these long wave-lengths. I have seen in preparation maps from Kitt Peak, by Rather and Ade, at 10 cm^{-1}, and from Hawaii, by the QMC group, at 25-30 cm^{-1}, which indicate features; the former, in particular, show the thermal contours at several phases moving relative to fixed features. If one was working with a high-resolution telescope, an average over several adjacent near-equatorial scans would be needed to eliminate such features, alternatively their effects could be included in calibration maps. The inclusion of that order of detail is not within the scope of the problem I set out to tackle here, but would, I agree, be a logical extension of this calibration proposal.

Martin Would the moon not be relevant for monitoring rapid fluctuations in atmospheric attenuation?

Kleinmann Yes, provided the source is within 20-25° of the moon, and that both are being observed at less than 2 air masses. Also, the same selenographic position should be used.

Traub There are efficient computer techniques (Grant J.P. and Hunt G.E., 1968, Mon.Not.R.ast.Soc. 141, 27) which are capable of determining multiple, not just single, scattering in inhomogeneous (discrete layers) media with arbitrary phase functions, using matrix solutions of the radiative transfer equation. Since these are relatively efficient and flexible calculating procedures, it seems a pity to use large amounts of computer time with any lesser technique.

Pugh At the time I encoded the program, I was not aware of Grant and Hunt's techniques. If I had been, I might still have used the simple approach given here, as it appears to be much faster to encode, although slower to execute, and I was therefore able to use up a surplus of computer allocation in the few days before the end of an allocation period.

PART 2

SOLAR AND JOVIAN ATMOSPHERES

THE FAR INFRARED SPECTRUM OF JUPITER

I. Furniss, R. E. Jennings and K. J. King

Department of Physics and Astronomy, University College, London W.C.1

Abstract   Far infrared spectra of Jupiter in the range 60-220 cm^{-1} are presented and compared with theoretical models.

1. Introduction

Knowledge of the atmospheric structure of Jupiter can be obtained from observations of the planet's far infrared emission. Measurements in the spectral range 60 to 220 cm^{-1} were made using a Michelson Interferometer on the University College London 40 cm Balloon-borne Telescope System (Tomlinson et al. 1974). The spectrum obtained is presented in section 3 and discussed with particular regard to the Jovian atmospheric models computed by Hogan et al. (1969) and the spectra predicted by Encranaz et al. (1971) using these models.

The interpretation of the emission spectrum of Jupiter in this region avoids the problems of scattering in the atmosphere at shorter wavelengths and the difficulty of separating the small thermal emission from the large non-thermal component arising in the Jovian radiation belts at longer wavelengths.

2. Instrument

The interferometer used is similar to that flown previously (Alvarez, 1974) with an 8.9 micron melinex beamsplitter, except that the present version employs two corner-cube reflectors. The f/6.8 beam from the telescope, after passing through the interferometer was focussed by a T.P.X.

lens, acting as the entrance window of the liquid helium cryostat, onto the
field of view aperture. The spectral range was defined by a black polyethylene filter and the quartz Fabry lens in front of the gallium-doped
germanium bolometer. This arrangement gave a reasonably flat response over
the frequency range 60 to 220 cm^{-1}. The field of view on the sky was 6' arc
and by moving the Cassegrain mirror of the telescope, using a solenoid system
driven by a 16 Hz square wave, the beam was switched between two points
6.3' arc apart. The phase sensitive detection of the signal was performed
on the platform and the resulting signal digitised before being telemetered
to the ground.

3. Observations

The observations were made during a flight on the night of 18th October 1974
at an altitude of approximately 30 km. The spectra were obtained in a total
of 65 minutes observation time, spectra being taken at apodised resolutions
of 0.5 cm^{-1} and 2.0 cm^{-1}. These have been degraded to a final resolution
of 4.0 cm^{-1} in order to improve the signal to noise ratio.

The pointing stability of the platform during the flight was good apart
from a small oscillation which occurred at a discrete frequency. The
corresponding modulation of the flux from Jupiter resulted in the spectrum
having a large feature at approximately 110 cm^{-1}. For this reason the region
105 to 115 cm^{-1} has been omitted from the spectrum. This modulation may also
account for the apparent loss of intensity above 200 cm^{-1} as the spectrum will
also contain a small 'ghost' spectrum between 170 and 330 cm^{-1} which may be
out of phase with the main spectrum.

Throughout the flight, spectra were also taken of the background signal which
arises from a small mismatch of the two beams in the system. These spectra
were taken at approximately the same elevation as the Jupiter spectra and had
about 50% of their intensity. It was found that the level of this background signal varied by rather more than expected giving rise to a systematic
error. The error bars shown in Figs. 1 and 2 are larger than the spread of
the points in the spectrum would indicate as a result of this. For this
reason the spectrum is plotted in Fig. 3 without error bars and the ammonia

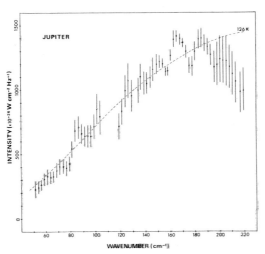

Fig. 1 Jovian **emission spectrum**. Resolution is 4 cm^{-1}.

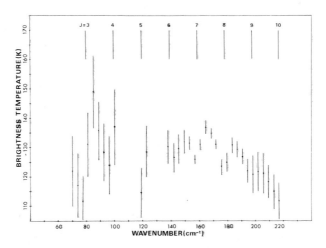

Fig. 2 Jovian brightness temperature spectrum. Solid lines above
spectrum show expected positions of ammonia absorption bands.

absorption bands can be seen clearly to be at the correct frequencies.

Low resolution background spectra were used to define the spectral
transmission of the system by assuming them to arise from emission of a
blackbody at the temperature of the telescope. This has been shown to be
the case in earlier flights (Alvarez, 1974). No correction has been made
for absorption in the atmosphere above 30 km because at the resolution used
no significant features should be seen.

As no absolute calibration source was included, a weighted least squares fit
to a pure blackbody curve was calculated and a colour temperature of 126 K
obtained. The spectrum values were then converted to brightness temperatures
showing more clearly the small deviations in the spectrum about the mean
temperature. To obtain absolute intensities, the area under the curve shown
in Fig.1 was equated to the expected integrated flux from a blackbody at that
colour temperature, with the size and at the distance of Jupiter.

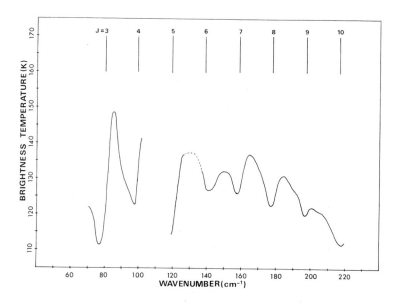

Fig. 3 Jovian brightness temperature spectrum. Solid lines above
spectrum show expected positions of ammonia absorption features.
The dashed portion of the curve is not well determined due to
absorption in this region by the quartz Fabry lens.

4. Discussion

A comparison is now made of the measured spectrum with that predicted.

The thermal profiles given by Hogan et al (1969) (see Fig.4) have been used by Encranaz et al. (1971) to calculate the emission spectrum of Jupiter in the 10 to 250 cm^{-1} region. The result using model 3 is shown in Fig.5. It is assumed that the measured brightness temperature at any frequency represents the temperature of the atmospheric level at which the optical depth in the atmosphere, at that frequency, is approximately 2/3. Between the ammonia absorption bands, the emission is expected to come from layers deep in the atmosphere below the inversion layer. Nearer the band centre the optical depth is greater and the emission comes from higher layers at lower temperatures. In the centre of the bands the absorption is very large and the emission comes from layers above the inversion level, which are at higher temperatures, giving rise to the emission features at the centre of each band.

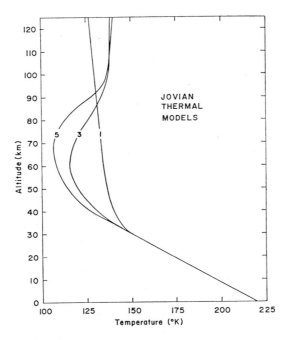

Fig. 4 Atmospheric thermal models computed by Hogan et al.(1969). Reproduced from Encranaz et al. (1971).

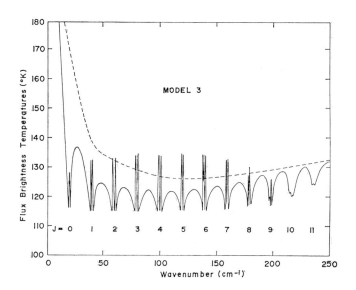

<u>Fig. 5</u> Predicted far infrared spectrum of Jupiter corresponding to model 3. Reproduced from Encranaz et al. (1971).

These emission features would not be seen at 4 cm^{-1} resolution. However, some spectra were obtained at 0.5 cm^{-1} resolution and allowing for the low signal to noise ratio no emission features were seen (The high resolution spectrum of a typical band (J = 7) is shown in Fig. 6. Again the errors are thought to be smaller than shown as all the features in the spectrum can be associated with water vapour features in this region.) The lack of these emission features can be interpreted in two ways;   the atmosphere does not have an inversion, in which case the temperature profile is similar to that shown in model 1 (see Fig. 4). However, this would contradict observations at shorter wavelengths (e.g. Gillett et al. 1969). The other interpretation is that the abundance of ammonia above the inversion level is less than that predicted by Encranaz et al.   This latter possibility is in agreement with the fact that the general level of the spectrum between the ammonia bands is approximately 135 K, higher than that predicted by Encranaz et al., but agreeing with broadband photometric measurements (Armstrong et al. 1972): Therefore the emission is coming from levels lower in the atmosphere and the abundance of ammonia above the inversion layer must be less than 1.0 cm atm. Aitken and Jones (1972) have measured a value of 0.22 cm atm. between 8 and 10 μm.

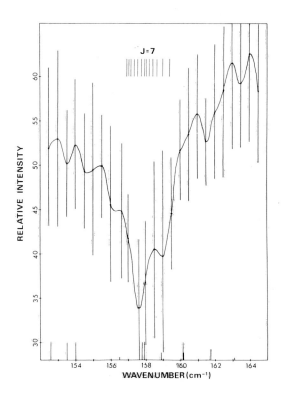

Fig. 6 High resolution spectrum of J = 7 absorption band. Solid
lines above spectrum show the expected positions of the ammonia
emission features. Solid lines below the spectrum show positions
of telluric water vapour absorption lines.

The minimum temperature observed in the band centres is approximately 120 K.
Recent shortwavelength measurements indicate the minimum to be at a
temperature of 118 K (Lacy et al. 1975).

Thus within the limited accuracy of these measurements, the interpretation of
the data is in general agreement with models deduced from near infrared
measurements.

Acknowledgments

We wish to acknowledge the assistance of W. A. Towlson, F. Want,

L. W. Britton and P. A. Roberts throughout the flight campaign and the personnel of the N.S.B.F. in Palestine, Texas, who were responsible for launch and recovery operations.
K. J. King holds an S.R.C. Studentship and I. Furniss is an S.R.C. Research Assistant.
This work was supported in part by an S.R.C. grant.

References

Aitken, D.K., and Jones, B., 1972, Nature, 240, 230.
Alvarez, J.A., 1974, Ph.D. thesis, University of London.
Armstrong, K.R., Harper, D.A., and Low, F.J., 1972, Astrophys. J., 178, L89.
Encranaz, Th., Gautier, D., Vapillon, L., and Verdet, J.P., 1971, Astron. Astrophys., 11, 431.
Gillett,F.C., Low, F.J., and Stein, W.A., 1969, Astrophys. J., 157, 925.
Hogan, J., Rasool, S.I., and Encranaz, Th., 1969, J.of Atm.Sci., 26, 898.
Lacy, J.H., Larrabee, A.I., Wollman, E.R., Geballe, T.R., Townes, C.H., Bregman, J.D., and Rank, D.M., 1975, Astrophys. J., 198, L145.
Tomlinson, H.S., Towlson, W.A., and Venis, T.E., 1974, Proceedings of the Symposium on 'Telescope Systems for Balloon-Borne Research' NASA TM X-62, 397.

DISCUSSION

Fazio What was the angular oscillation of the telescope compared to the spectrometer beamwidth? Was the oscillation at a definite frequency or was it random?

King The field of view of the telescope was 6' arc and the oscillation amplitude was approximately 1.5' arc peak-to-peak. The oscillation appeared to be composed of just a few specific frequencies, only one of which came within our frequency range.

Carli How did you calculate the noise in your spectra?

King The error bars correspond to one standard deviation of the individual Jovian spectra about the arithmetic mean of those spectra.

Traub Current model atmospheres for Jupiter by Wallace,L., Prather, M., and Belton, M.J.S., (1974, Astrophys. J. 193, 481) show a very shallow inversion over a large amplitude range. Since the NH_3 mixing ratio drops off rapidly above the saturation level, one is unlikely to ever see any NH_3 emission cores, at least not on the scale suggested by Hogan et al. (1969).

NEW PARAMETERS FOR SOLAR SPICULES BASED ON SUBMILLIMETRE DATA

J. E. Beckman

Astronomy Division, Space Science Department,
European Space Agency, Noordwijk, Holland

and

J. Ross

Queen Mary College, London E.1

Abstract New sub-millimetre observations of the quiet solar limb
with angular resolution 1.5" arc show a narrow asymmetric
brightness spike centred near the optical limb, with approximate
half-width 10" arc. We present a semi-empirical predictive model
based on the properties of an individual spicule derived from
optical and UV data. From it one can derive electron densities
and temperatures within a spicule, a scale height of 2000 km for
groups of spicules, and a mean number density over the solar
surface of $4.2 \ 10^{-8} \ km^{-2}$ at 1,200 km height above the photosphere.

1. Experiment and Data.

The equipment flown during the total eclipse of June 30th 1973 was designed
to measure the sub-millimetre output of the sun's limb as a function of time
during 2nd and 3rd contacts. This is then transposable to yield a plot of
brightness temperature with angular distance from the limb. Fig. 1 shows how
radiation from the sun-moon disc was taken in through a crystal quartz window
and passed by a series of plane mirrors MI to M5 and a condenser LI into a
rapidly scanned Michelson interferometer. The moving mirror M6 could be
oscillated to yield an interferogram every 1 sec or 10 seconds. The
interferograms produced by the beams from M6 and M7 were focused onto a
helium-cooled Rollin InSb detector, and the output from the detector was
stored on analog tape. Subsequent digitization and Fourier transformation

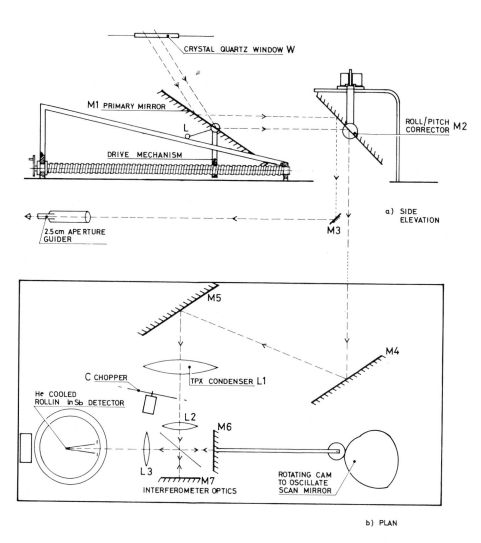

CRYSTAL QUARTZ WINDOW W

M1 PRIMARY MIRROR

L

DRIVE MECHANISM

ROLL/PITCH
CORRECTOR M2

a) SIDE
ELEVATION

M3

2.5 cm APERTURE
GUIDER

M5

C CHOPPER

TPX CONDENSER L1

He COOLED
ROLLIN In Sb DETECTOR

L2

M6

M4

L3

M7

INTERFEROMETER OPTICS

ROTATING CAM
TO OSCILLATE
SCAN MIRROR

b) PLAN

Fig. 1.   The experimental setup for the eclipse measurements.

gave an output signal which was later split for good signal to noise into
three broad bands weighted at 400 μm, 800 μm and 1200 μm wavelengths. Fig.2
shows, for the case of 3rd contact, how the signals grew with time as the
solar limb emerged. The two underlying linear curves show what would have
been seen if the sun had been radiating as a uniformly bright disc, having
either its optical radius or a radius 5' arc greater than that. The upper
curves show the data in each channel, and indicate that the sub-millimetre
emission not only extends to 5' arc beyond the optical limb, but that the
limb is much brighter than the disc over a narrow angular range. Fig.3
illustrates this very strikingly, showing the derived brightness spike in
terms of angular distribution. The absolute temperature scale relies on an
extrapolation to the disc centre values for the brightness temperature T_B
at each wavelength, and is not reliable to within 10%. However, the spike
shape is well shown, and strong conclusions can be drawn from its near
invariance with wavelengths between 1200 μm and 400 μm. Finally, in Fig.4,
we present a more detailed plot of the data at 1200 μm. Here the
observational error bars are indicated and the T_B scale has been arranged to
give a disc centre value of 5700 K, in agreement with previous work[1,2].

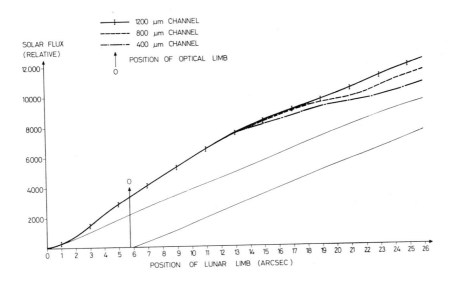

Fig.2 The growth with time of the signal as the solar limb
       emerged, at 3rd contact.

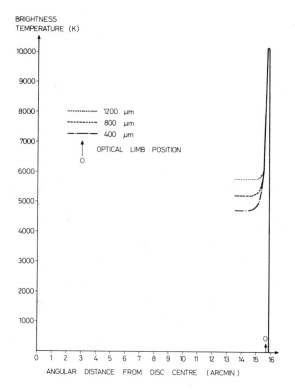

Fig. 3. The derived brightness spike in terms of angular
         distribution.

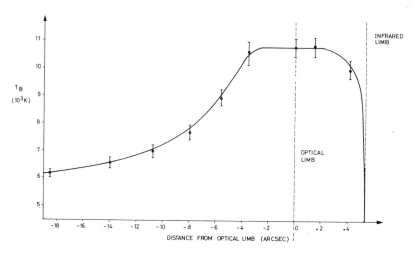

Fig. 4. A more detailed plot of the 1200 μm data.

The solid curve is the value predicted by a theoretical model described
below. Fig.4 illustrates how restricted was the portion of the sun
observed: the solar disc at the time of observation had a radius of 944" arc.

2. General Properties of a Spicule Model

Combining the shapes of the inner and outer curves of the limb spike can give
new information about spicules, given various physical assumptions. The main
assumption is that spicules are responsible for the spike. We will not show
here that alternative models, using coronal sheets, do not give a good fit to
the data (which can be demonstrated) but will assume a spicular model, and
show that it can fit the data well. A second assumption is that the
chromosphere can be represented globally by an optically rough emitter,
through which the spicules protrude. Such a model has been invoked[3,4] to
account for the observed absence of macroscopic sub-millimetric limb
brightening, on an arcminute scale. The scale length for the roughening is
1500 km, and the radiative floor thus provided varies in height h above the
photosphere between h = 800 km at λ= 400μm, and h = 1500 km at 1200μm. For
comparison the chromopause, the kinetic temperature minimum, lies at
approximately 500 km above the photosphere, equivalent to 0.7" arc above the
optical limb. Linear distances rather than optical are employed, because the
chromosphere is highly inhomogeneous and non-isotropic.

The sub-millimetre emissive properties of an individual spicule are derivable
from the work of Beckers[5,6] and are tabulated below:-

h(km)	T_e	$N_e (cm^{-3})$	T_{eff} (1200μm)	T_{eff} (800μm)	T_{eff} (400μm)	T_{eff} (100μm)
1,000	8,000	1.65×10^{11}	8,000	7,500	5,500	350
2,000	9,000	1.6×10^{11}	9,000	8,500	5,000	200
4,000	12,000	1.5×10^{11}	11,750	8,500	4,000	100
6,000	14,000	8.9×10^{10}	12,000	8,000	2,500	25
10,000	16,000	3.4×10^{10}	9,500	5,750	1,500	-

At 1200μm wavelength we can treat each spicule as an optically thick emitter

out to heights of 6000 km, corresponding to 8" arc in projected length, with
an effective brightness temperature varying between 8000 K and 12000 K, then
falling away above 10,000 km.   The observation that the 400 μm wavelength
curve has virtually the same shape as the 1200μm curve within the limb
(outside our angular resolution is rather too coarse as a fraction of scale
height) shows that Becker's electron density estimates below 4000 km must be
too low by a factor of 1.5.   However, the general form of radiance
distribution expected from an array of spicules each having physical
properties is in fair agreement with those optically derived.

3.   The Spicule Distribution.

Optical observations[7,8] show that above   h = 3000 km the number distribution
of spicules is well represented by

$$N(h) \quad = \quad N_o e^{-\alpha h} \tag{1}$$

where N_o is the number per unit area at a reference level at or close to the
photosphere, and α is the inverse scale height for the spicules.   Since at
1200 μm each spicule is optically thick, the observed spike is a tangential
observation of the superposed temperatures of individual spicules.   In
particular, outside the limb the fractional area covered by spicule material
should be represented by

$$N(h) \quad = \quad N(h) \left\{ 1 + \frac{1}{(\alpha h)^2} \right\} \tag{2}$$

to a first approximation, with $\left\{ 1 + \frac{1}{(\alpha h)^2} \right\}$ representing the solar limb
geometry at the essentially cylindrical edge.   The best fit of this outer
shape to the data yields a value for α   of 2000 (± 200) km, a result which is
independent of the value assumed for N_o.   The portion of curve used to give
the fit lies near   h = 3000 km, so that the effect of overlapping is not large,
and the geometry factor is also less than 30%, so that approximative errors
are quite small.

Within the limb, overlapping has to be taken into account.   Given a random

array of vertical spicules having a scale height α, and mean diameter b, the
fractional coverage of projected surface $S(\phi)$ as a function of angle ϕ,
where $\pi/2 - \phi$ is the angle between the observer's line of sight and the
vertical at a point on the surface, is given by

$$S(\phi) \quad = \quad 1 - e^{-N_o b\alpha \cot \phi} \qquad\qquad (3)$$

and for our derived value of α, and an optical value of b of 850 km we
obtain for N_o, the value of 4×10^{-8} km^{-2}.   These values correspond to a
number over the whole solar surface of 250,000 spicules, and at h = 3000 km
a value of 55,000, in good agreement with optical observations. Our data thus
enables the observer to go lower than the 3000 km limit set by the Hα and
He lines and the Lyman and Balmer continua, which are not so well-
determined in their dependence on kinetic temperature.

4. Conclusions

The analysis presented shows that the observations obtained[9] with angular
resolution 1.5 arcseconds (more than an order of magnitude finer than
previously available) can be well interpreted in terms of a spicule model.
The model demands an increase in electron density within a spicule over that
from optical spectral data by a factor 1.5 at heights h less than 4000 km
above the photosphere. With this change incorporated, the number densities of
spicules agree well with those obtained by other workers using Hα
chromospheric spectroheliograms, over the range of heights where the two types
of observations overlap. One can thus have confidence that our predicted
number distribution for lower values of h is a good one. A relatively
clear-cut observational test is possible; in the present model spicules
should become optically thin between 400μm and 100μm wavelength. Thus an
extension of the present measurements to 100μm wavelength and below would
strengthen the model, and at the same time yield more accurate values for
electron number densities within the spicules. The radiometry would be
further improved by extending the measurements well within the solar limb.

Acknowledgments

The authors acknowledge financial and technical help from the S.R.C. and the
Appleton Laboratory.   The flight was made possible by the I.N.A.G.

department of the French Government's D.G.S.T., with technical assistance
from the personnel of S.N.I.A.S. led by M.A. Turcat.   One of us, (J.R.)
is in receipt of an S.R.C. research studentship.

References

1. Fedoseev, L.I., Lubyako, L.V., and Kukin, L.M.; Soviet Astron A.J.
                                              11, 953, 1968.
2. Linsky, J.L.; Solar Physics 28, 409, 1973.
3. Beckman, J.E., Clark, C.D., and Ross, J.; Solar Physics 31, 319, 1973.
4. Kalaghan, P.M.; Solar Physics 39, 315, 1974.
5. Beckers, J.M.; Solar Physics 3, 367, 1968.
6. Beckers, J.M.; Ann.Rev. Astr. and Astrophys. 10, 73, 1972.
7. Rush, J.H. and Roberts, W.O.; Aust.J.Phys. 7, 230, 1954.
8. Athay, R.G., Astrophys.J. 129, 164, 1959.
9. Beckman, J.E., Lesurf, J.C.G., and Ross, J.; Nature 254, 38, 1975.

DISCUSSION

Coron   Is it possible that the spike at the limb is polarized?

Beckman   We were not in a position to measure polarization. The spicule
model does not predict a high degree of polarization. I am not sure about
the coronal model, but a first thought suggests that will not give strong
polarization either.

SOLAR RADIOMETRY IN THE MILLIMETRE REGION WITH A LOCAL CALIBRATION SOURCE

B. Carli

Queen Mary College, London E.1

Abstract A system for solar radiometry in the millimetre region, as an alternative to calibrations using the moon, is described. The effects of long wavelength diffraction and atmospheric absorption are discussed.

1. Introduction

Because of the varying opacity of the solar atmosphere, measurements of the solar brightness temperature as a function of frequency can be used to study the different layers of the solar atmosphere. The submillimetre and millimetre emission has the great advantage of being characterised by a smooth and well known opacity (H^- free-free), and precise radiometric measurements of the solar temperature in this region provide important information for the study of the photosphere and the low chromosphere of the sun. But the available observations suffer a relatively large error bar due to calibration uncertainties and are at the moment mainly used for consistency check on models based on visible and ultraviolet measurements.

In this paper we examine the sources of error in the millimetre calibration and consider the possibility of using a local calibration to improve the precision.

2. Calibration using the moon

The moon is generally used as the reference source for the measurement of the solar brightness temperature, the main advantage being that a relative

calibration with a source external to the atmosphere does not require
corrections for the atmospheric transparency.

The precision of this measurement is based mainly on the precision with
which the moon is known as a radiometric standard. The brightness
temperature of the moon in the millimetre region is a function of many of
the lunar physical properties. It depends on the opacity of the lunar soil
as a function of wavelength and temperature, on the temperature of the lunar
surface as a function of the depth and the lunation time, and, as
demonstrated by M.Pugh (p.63 , this volume) on the scattering properties of
the lunar soil.

Two temperatures are usually considered: the new moon brightness
temperature and the brightness temperature averaged over the lunation period.
The errors on the lunar physical properties affect directly the calculation
of the new moon temperature, whereas some of these errors enter only as a
second order effect in the temperature averaged over a lunation period. For
this reason the average temperature is frequently quoted in literature.
Linsky[1], for instance, has calculated this quantity with an error of only
about 4%. Measurements are continually improved and more precise models
developed, but the error quoted by Linsky can be taken as the order of the
error one should expect in a calibration done using the moon as a reference.
This error is not critical for the calibration of faint sources which are
acquired with a relatively small signal-to-noise ratio, but are the limiting
factor in solar radiometry.

3. Calibration using a local source

A new approach to the problem of solar radiometry is here tried using the
polarizing interferometer[2] and a local calibration source. With the
interferometer it is possible to measure simultaneously several spectral
bands, and the local source overcomes the limitations of the moon
calibration. In the following, the three main difficulties presented by
this technique are discussed.

(a) Black-body To build a cavity with high absorption at wavelength around
one millimetre, it is necessary to use either a thick absorbing material,

which is difficult to maintain at a uniform temperature, or a very large
cavity. A black-body cavity which is a useful compromise has been recently
developed. It uses a very thin absorbing material (0.1 mm), is small enough
to be handled by one person, and has a reflectivity smaller than 0.001[3]

(b) Diffraction Because of diffraction, the absolute calibration of the
interferometer mounted on a telescope is very complex. For this reason we
have chosen a very simple optical system which allows a direct calibration.
The sun is followed with a flat mirror which rotates on a heliostat in a polar
mounting. The calibration is done replacing the flat mirror with a concave
one which has the black-body in its focus. Both mirrors are overdimensioned
so that no diffraction is introduced and both the sun and the black-body are
seen by the interferometer as two sources in front of it, at infinity. The
absence of optics compels one to observe the whole solar disc rather than a
defined and quiet region in the centre of the sun. But the same information
is obtained if the calibration is done in parallel with a map of the sun.

The signal coming from the sky around the sun and from the ambient around the
black-body is in both cases internally suppressed using the two input ports of
the interferometer[2].

(c) Atmospheric transparency Uncertainties in the atmospheric transparency
are the main source of error in this measurement. Nevertheless, their effect
can be minimized by taking measurements in very dry conditions from a high
altitude observatory or, more drastically, from an aircraft or balloon.

4. Data analysis

The measured signal $S(\lambda)$, normalised relatively to the black-body
calibration $C(\lambda)$, is equal to:

$$\frac{S(\lambda)}{C(\lambda)} = T_\theta(\lambda)\, e^{-k(\lambda)w}$$

where

$T_\Theta (\lambda)$ is the solar brightness temperature,

$e^{-k(\lambda)w}$ is the atmospheric transparency,

$k(\lambda)$ is the water vapour absorption coefficient and

w is the amount of water vapour in the line of sight.

If we express the spectral dependency of the sun temperature in the following way[4,5]:

$$T_\Theta (\lambda) = T_\Theta (\lambda_o)(\lambda/\lambda_o)^\alpha$$

we obtain

$$\frac{S(\lambda)}{C(\lambda)} = T_\Theta(\lambda_o) (\lambda/\lambda_o)^\alpha e^{-k(\lambda)w}$$

and from the fitting of the experimental data with a regression curve, the values of T_Θ, (λ), α and w are obtained.

This technique was tried for the first time during a site testing programme at the Observatory of Gornergrat (Switzerland) in September 1974. The experiment was done in collaboration between Queen Mary College and the University of Florence. The sun was observed for three days but, unfortunately, in none of these was the transparency high, the minimum precipitable water vapour content in the line of sight being 3 mm. The spectra were studied in the frequency range 4.5 - 13.5 cm^{-1}. The variation of the solar brightness temperature with frequency was successfully separated from the large atmospheric effects. Some systematic differences are present between the experimental data and the regression curve, and stress the lack of a sufficiently detailed model for a rigorous correction of the atmospheric absorption when working with high water vapour concentrations.

A value of $\alpha = 0.4$ was obtained. It is somewhat higher than expected. This could be due to the fact that we observe the full solar disc, and limb effects (see Beckman and Ross, p.79 this volume) and solar flares (see Croom, p. 93 , this volume) also contribute to our signal. A more detailed data analysis is to follow.

5. Conclusions

Reliable black-body sources in the millimetre region make it possible to overcome the limitations of the calibration with the moon and to measure the solar brightness temperature with increased precision. A system for solar radiometry with a local calibration source was studied. In the case of high water vapour content, the system is limited by the fact that the atmospheric absorption cannot be satisfactorily corrected for. Nevertheless, first measurements, taken with relatively poor atmospheric transparency, give consistent results.

References

1. Linsky, J.L., 1973, Astrophys. J.Suppl. Series No.216, 25, 163-204.
2. Martin, D.H., 1972, in Infrared Detection Techniques for Space Research, V.Manno and J.Ring(ed.),(D.Reidel, Holland).
3. Carli, B., 1974, IEEE Trans. MTT 22, 1094.
4. Linsky, J.L., 1973, Solar Physics 28, 409-418.
5. Beckman, J.E., Clark, C.D. and Ross, J., 1973, Solar Physics 31, 319-338.

DISCUSSION

Beckman Did you measure the sun's absolute brightness temperature at any particular wavelength?

Carli The responsivity of our detector was not constant during the measurements and we have lost in this first experiment the information on the absolute temperature of the sun at our particular frequency.

Marsden What type of blackbody source did you use for calibration? What temperature has it been used at?

Carli The blackbody we have used can be described as an off-axis conical cavity. The geometry is such as to maximise the number of reflections for perpendicularly incident radiation. The absorbing surface is a dielectric coated with a resistive metal film. We actually used two blackbodies: one was heated up to 140°C and the other was cooled to liquid nitrogen temperature.

Mercer Does the blackness of the blackbody you referred to become less black at angles away from the optic axis?

Carli Yes, it is black within an angle of $\pm6°$. The number of reflections in the cavity and the efficiency of the cavity decrease rapidly outside this angle. This behaviour is very desirable because it minimizes the heat input from directions which are not relevant.

SOLAR FLARE OBSERVATIONS AT MILLIMETRE AND SUB MILLIMETRE WAVELENGTHS

D. L. Croom

S.R.C., Appleton Laboratory, Ditton Park, Slough, SL3 9JX, Berks, U.K.

Abstract Observations of solar flares at millimetre wavelengths during the sunspot cycle now ending have shown that some flares emit the bulk of their radio energy in the mm band. The problems of extending such observations to the sub-millimetre region are considered.

1. Introduction

The objective of this paper is to summarise the main features of millimetre-wavelength solar flare observations and to attempt to extrapolate the available data to the far infra-red region (wavelength $\lambda \leq 1$ mm). In doing this it must be borne in mind that because both the time of occurrence of the flare and its exact location on the solar disc are not known beforehand, and because the flares are relatively short-lived and contain fine time-structures, the most efficient way of obtaining data is to use a telescope beamwidth of at least 1°, or preferably more, so that the whole solar disc can be monitored continuously with approximately equal weighting being given to every point on the disc. Microwave/millimetre wave bursts have been observed with large, narrow-beam telescopes operating in a scanning mode. Although this enables much weaker events to be recorded, many are inevitably missed, and those that are recorded usually lose much of their detail. In particular the peak intensity is often lost with this technique. In addition, large telescopes are not normally available for continuous long-term solar observations.

2. Millimetre Wavelength Observations of Solar Flares

Prior to the beginning of the current sunspot cycle (20th) virtually all solar flare studies in the radio region of the electromagnetic spectrum were carried out at wavelengths longer than 30 mm, with the main emphasis being

in the wavelength region above 300 mm (decimetre and metre bands). With
one exception, such observations as were carried out at wavelengths shorter
than about 30 mm resulted in the detection of only minor events. (Piddington
and Minnett 1949; Hagen and Hepburn 1952; Coates 1958). The one exception
was a recording made by Coates (1966) of an intense burst at 4.3 mm wave-
length having a peak flux increase of about 5400×10^{-22} W m^{-2} Hz^{-1} = 5400 sfu
(sfu = solar flux unit = 10^{-22} W m^{-2} Hz^{-1}). The intensity of this event can
be judged by the fact that any microwave event whose peak flux increase is
≥ 500 sfu (less than a tenth of the value reported by Coates) is classified
as a major event (only about 5% of all microwave bursts come into this
category). However, in spite of this, and of the fact that many burst
spectra, particularly those associated with proton events, were still
increasing with decreasing wavelength at 30 mm, no further shorter wavelength flare studies were carried out until the 20th sunspot cycle in the
mid 1960's when it was decided at Appleton Laboratory, Slough, to carry out
an extensive study of solar flares at wavelengths, of 4.2, 8.1 and 15.8 mm
(arrowed in Fig. 2) together with longer wavelengths. In addition Air Force
Cambridge Research Laboratories (AFCRL), U.S.A., added observations at
8.6 mm, to their longer wavelength (cm and dm) observations. The AFCRL
and Appleton Laboratories have a significant overlap in observing times, and
therefore wherever possible published data from the two observatories have
been combined to produce the spectra shown in this paper.

3. Microwave-millimetre Wave Flare Spectra

Fig. 1 illustrates in sketch form the microwave spectrum classification
of Castelli et al. (1969), and is based almost entirely on the direction of
the spectral slope, for wavelengths longer than 30 mm. These classes can
be further subdivided according to intensity.

Types A and G can occur separately or jointly. When they occur jointly,
and when the observations are restricted in wavelength range then one records
what has become known as a U-shaped spectrum, (Fig. 2), which when coupled
with intensity criteria has been used successfully by Castelli et al. (1967)
as an indicator of solar proton events.

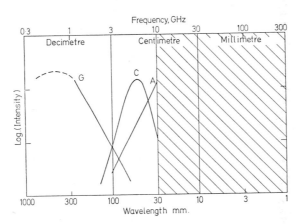

Fig. 1. The burst spectrum classification of Castelli et al. (1969).

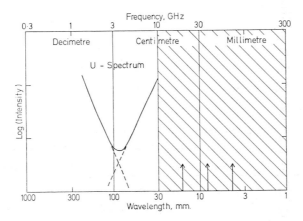

Fig. 2. The U-shaped spectrum of Castelli et al. (1967). The arrows indicate the mm wavelengths used by Appleton Laboratory (AFCRL observe at 19.4 and 8.6 mm).

are extended down to 4 mm in wavelengths, (and includes the dm wavelength spectrum in full). The general picture of the radio flare event that emerges is of a decimetre wavelength event due to electrons ejected outwards from the source into the corona, and a cm-mm event due to electrons ejected downwards from the source into the chromosphere. That these are distinctly separate events is borne out by the morphology of the bursts which is generally significantly different for the two wavelength regions. The type

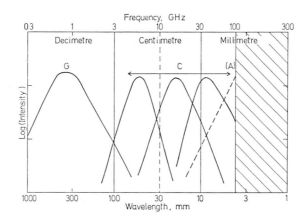

<u>Fig. 3.</u> Idealised form of burst spectra when observations down
to 4 mm wavelength are incorporated.

A event, when observations are extended to 4 mm, virtually disappears, and
hence it is shown as a broken line in Fig. 3. Only one event still rising at
4.2 mm has so far been recorded, and the spectrum of this event almost
certainly turns over between 3 and 4 mm.

However, in addition to the dm-microwave combinations we have found a small
number of cases in which the spectra have separate cm and mm peaks which
would seem to imply either two separate particle streams ejected into the
chromosphere, or one particle stream producing radio bursts by two separate
mechanisms. The latter suggestion is supported by the fact that there is
no appreciable difference in burst morphology between the cm and mm peaks
(though there is between the cm and dm events). Thus we come to a range
of radio spectra as idealised in Fig. 4. Examples of real spectra corresponding to Fig. 4. can be found in Croom (1972).

In considering now the extension of burst observations into the far infra-red
region it is necessary to emphasise two points that emergy from a statistical
study by Castelli and Guidice (1972).

(a) Most (approx. 76%) microwave bursts peak in the 60-30 mm wavelength
     range.

(b) Although only 5% of all microwave bursts exceed 500 sfu those that
     do so can reach 100 times this value.

Nevertheless, many of the microwave events whose spectral maximum lies at

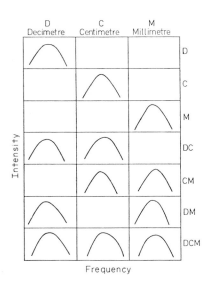

Fig. 4. Idealised spectral classification of solar flare events
in the decimetre to millimetre wavelength region.

longer than 30 mm, need data in the shorter wavelength region to define the
spectral peak and the slope of the high frequency cut-off, both of which are
important factors in any discussion of the radiation mechanism involved.   In
particular the main part of the radio energy of the event can occur in the
mm region, even though the peak emission is in the cm region (Croom 1971).

Further, those events which peak in the mm region, although relatively
few in number, are usually associated with the more important solar events,
particularly in relation to their terrestrial effects, and can be as intense
as any recorded at cm wavelengths (Croom 1970).

Fig. 5 shows some spectra of actual events, from which it can be seen that
some at least, of these flare events should be detectable at 1 mm or less.
There is however an observational problem involved connected with the
discussion in section 1 concerning whole-disc monitoring versus solar
mapping. This is illustrated in Fig. 6, which shows (curve a) the increase
in background solar flux as a function of decreasing wavelength (based on

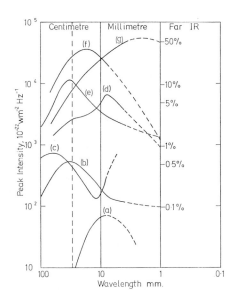

Fig. 5. Examples of burst spectra, with indication of plausible extrapolations to 1 mm. The percentage scale relates to the quiet sun flux at 1 mm.

(a) 21 Sept. 1969 (1756 UT)   (e) 20 July 1970 (1124 UT)
(b) 29 May 1970 (1127 UT)   (f) 23 May 1967 (1947 UT)
(c) 13 June 1970 (1233 UT)   (g) 6 July 1968 (0947 UT)
(d) 27 Feb. 1969 (1408 UT)

(Data from Appleton Laboratory, Slough, and AFCRL, Mass.)

the Bilderberg model of Gingerich and de Jager (1968), though the exact model is not critical). Curve b shows the effect of a flare increase of 500 sfu (the major burst threshold) and curve c the effect of a burst reaching what must, on observational evidence to date, be regarded as the maximum enhancement likely to be obtained at any cm-mm wavelength (occurrence rate of the order of once or twice per solar cycle).

Assuming that, under ideal atmospheric conditions (a rarity at mm wavelengths) an enhancement of 1% of the solar background is detectable, then at 1 mm wavelength only bursts about 1000 sfu will be detectable, and at 0.1 mm even the largest events (supposing that they peaked at that wavelength) would be undetectable.

Impulsive type events at cm-mm wavelengths are generally accepted as

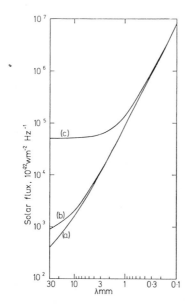

Fig. 6. The potential enhancement of solar flux due to flare
events.
(a) Quiet sun emission.
(b) Enhancement due to burst of 500 sfu peak increase.
(c) Enhancement due to burst of 50,000 sfu peak increase.
(Approximately a one per sunspot cycle event.)

resulting from a gyro-synchrotron mechanism due to sub-relativistic electrons,
although some of the larger ones could be produced by synchrotron radiation
from relativistic electrons (event e in Fig. 5 could be an example of the
latter, characterised by a low rate of short wavelength cut-off). For events
due to the gyro-synchrotron mechanism the frequency f_{max} at which the
spectrum peaks can be approximately related to the magnetic field H in the
source by (Takakura 1967):

$$f_{max} \underset{\sim}{} 3.5 \ f_H \underset{\sim}{} 9.8 \ H \qquad\qquad (1)$$

where f_H is the gyro-frequency, f is in MHz and H in gauss. Thus
for a gyro-synchrotron event to peak at a wavelength of less than 3 mm a
magnetic field exceeding 10,000 gauss would be required, and this is already
somewhat larger than the largest intensities thought to exist over active

regions. This is borne out by the fact that it is extremely rare for an
event to peak at a wavelength of less than 15 mm (H $\stackrel{\sim}{\sim}$ 2000 gauss), and
only 1 event has been observed with a peak at 4 mm (H $\stackrel{\sim}{\sim}$ 7200 gauss).

If relativistic electrons are involved, then f_{max} is also a function of
the energy of the electrons:

$$f_{max} = 4.7 \ H \ E^2 \tag{2}$$

where E is in Mev, and f and H as before. For this mechanism,
considerably smaller magnetic fields are necessary. For example, for 5 Mev
electrons, a magnetic field of only about 640 gauss is required for a
burst to peak at 4 mm, and about 2500 gauss for it to peak at 1 mm.
This mechanism probably accounts for the intense events with peaks at wavelengths of less than 10 mm.

To sum up, it would seem that the detection of flare events at wavelengths
of 1 mm and less would be very valuable in helping to establish the rate
of high frequency cut-off for intense solar events, particularly those of
type IV, but that even without atmospheric attenuation problems, this will
be difficult except perhaps for a very few events whose peaks are recorded
probably by chance by large antennas studying other solar phenomena. For
completeness it must be emphasised that other forms of solar activity can be
detected at 1 mm and below with large antennas, but are generally slowly-
varying enough for complete time information to be obtained (see for example
Clark and Park 1968). These are however, not solar flare events.

Acknowledgement

The work described above was carried out at the Science Research Council's
Appleton Laboratory, and is published with the permission of the Director.

References

Castelli, J. P., Aarons, J. and Michael, G. A., 1967, J.Geophys.Res., 72,
     p. 5491.
Castelli, J. P., Aarons, J., Michael, G. A., Jones, C. and Ko, H. C., 1969,
     in "Solar Flares and Space Research", (North-Holland, Amsterdam), p.194.
Castelli, J. P. and Guidice, D. A., 1972, AFCRL Environmental Research Paper,
     No.381, Hanscom Field, Bedford, Mass., U.S.A.

Clark, C. D. and Park, W. M., 1968, Nature, 219, 924.

Coates, R. J., 1958, Nature, 182, 861.

Coates, R. J., 1966, Proc.IEEE, 54, 471.

Croom, D. L., 1970, Solar Phys., 19, 414.

Croom, D. L., 1971, Nature (Phys.Sci.), 229, 142.

Croom, D. L., 1972, Proc. 2nd Meeting of Committee of European Solar Radio
    Astronomers (CESRA), Trieste Astronom.Observ.

Gingerich, O. and de Jager, C., 1968, Solar Phys., 3, 5.

Hagen, J. P. and Hepburn, N., 1952, Nature, 170, 244.

Piddington, J. H. and Minnett, H. C., 1949, Austr.J.Scient.Res.A, 2, 539.

DISCUSSION

Rengarajan Are the millimetre bursts always accompanied by X-ray bursts?

Croom Virtually always, though it might be possible to find a few events
that were not. As far as I know all flare events peaking in the millimetre
region have associated X-ray emission. However the reverse is not true, as
many radio events which are accompanied by X-rays have cut-off at longer
wavelengths.

Martin Are the decimetre, centimetre and millimetre events clearly separate -
due to distinct mechanisms?

Croom The decimetre events is clearly distinguishable from the centimetre-
millimetre event, and this is clearly seen in the burst morphology. The
decimetre event is generally accepted as being due to synchrotron radiation
from relativistic electrons ejected outwards from the flare. The centimetre
event is likewise in general probably due to gyro-synchrotron radiation
from sub-relativistic electrons ejected down into the chromosphere, though
some events, including the more intense millimetre events may result from
relativistic electrons. One way of distinguishing these mechanisms is by
estimating the rate of high frequency cut-off of the burst spectra, and this
is often difficult or impossible simply because of lack of observational
data in the mm region. In those cases where separate cm and mm spectral
peaks are observed, there is no obvious change in burst morphology, suggesting
that perhaps two mechanisms are operating in one electron stream. However
observations are entirely absent in the region between 15 and 8 mm, the
region where the cut-out reaches its minimum. We hope to remedy this at AL
for the next sunspot cycle.

Beckman The QMC group has observed several bursts at 1.2 mm associated with
violent lower chromospheric disturbances (Ellerman "bombs"), which have also
been observed optically.

Croom I should emphasise that I have been considering only flare-assocaited
events. As Dr. Beckman points out there are other forms of mm activity,
some of which are not detected at longer wavelengths.

PART 3

COSMIC MICROWAVE BACKGROUND

A LAMELLAR GRATING INTERFEROMETER EXPERIMENT TO DETERMINE THE SPECTRUM OF THE COSMIC BACKGROUND RADIATION

J. B. Mercer, S. Wilson, P. Chaloupka, W. K. Griffiths, P. Marchant,

P. L. Marsden and C. C. Morath

Physics Department, University of Leeds, LEEDS, England

Abstract: A balloon-borne interferometer experiment to measure the cosmic background radiation over the spectral range $3 \leq \tilde{\nu} \leq 12$ cm^{-1} is described. In addition, calculations on the expected atmospheric emission are presented.

1. Introduction

Because of the profound implications for cosmological theory, measurement of the spectrum of the cosmic background radiation is extremely important in the region of the spectral peak. Early balloon and rocket experiments detected a considerable excess of radiation but subsequent flights produced results not incompatible with a 2.7 K Planckian spectrum. These flights carried broad band experiments and so provided limited spectral information. However, two recent balloon flights[1,2] have made interferometric measurements of the cosmic background radiation.

We describe here a new experiment designed to trace out the spectrum near its expected maximum. The apparatus consists of a balloon-borne platform and a cryostat containing a lamellar grating interferometer with an InSb detector. Since the platform is steerable, coarse isotropy measurements can be made in the azimuthal plane.

We also present calculations of the expected contribution due to emission from the residual atmosphere (at ~ 4 mb) and point out the limitations these impose on balloon experiments of this type.

2. The Apparatus

2.1 The Optical Assembly

The main features of the optical system are shown in Fig.1. Radiation enters the optical system through a side window near the bottom of the cryostat. It is subsequently guided along a converging rectangular light pipe which condenses the beam cross section from (5×1.6) cm^2 to (2.2×0.5) cm^2, allowing it to be conveniently modulated by a rotating chopper. The succeeding light pipe diverges the beam. The emergent end of this light pipe serves as the entrance aperture for an off-axis paraboloid which collimates the beam onto the face of the lamellar grating interferometer. After reflection from the interferometer, the radiation is re-focussed by the paraboloid onto its exit aperture which is the mouth of a final converging light pipe. This converges the radiation onto the detector.

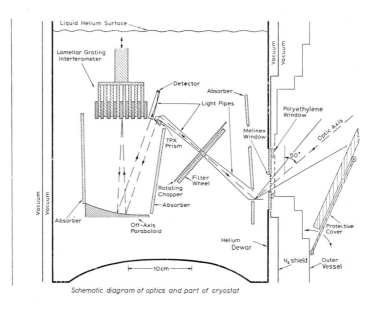

Schematic diagram of optics and part of cryostat

Fig. 1

The 'entrance' and 'exit' apertures are symmetrically displaced with respect
to the optic axis of the paraboloid.   These apertures have dimensions
0.8 cm × 5.0 cm parallel and perpendicular to the plane of the drawing
respectively. This geometry was chosen to maximize the étendue for the size
of lamellar grating selected (8.5 × 11.0 cm^2).   The resulting étendue,
~ 0.8 cm^2 sr, when conserved through the light pipe system, projects a broad
rectangular solid angle ∿$(8^o × 30^o)$ onto the sky.

The first two light pipes are fabricated from highly polished sheets of brass
while the third light pipe is electroformed from high purity copper.
Transmission measurements have been made on similar pipes and indicate an
average reflectivity of 0.98 in the relevant spectral region.   The transmission through the system, including the effects of gaps, corners, reflectance
and absorption losses, is calculated to be∿0.4.

The interferometer uses plates of Dural and has a grating constant of 1.0 cm.
This implies that the low frequency cut off is well under $\tilde{\nu}$ = 3 cm^{-1} while its
theoretical efficiency is 100% out to $\tilde{\nu}$ = 16 cm^{-1}, decreasing slowly there-
after. Maximum attainable resolution (unapodised) $\Delta \tilde{\nu}$ = 0.1 cm^{-1} is
determined by the maximum relative displacement of the plates (2.5 cm).   The
interferometer is driven by a stepping motor housed at the top of the cryostat.

2.2 Cryostat

The optical system is completely immersed in liquid helium in order to prevent
the detection of spurious effects due to temperature instability. The
cryostat consists of an 80 l.capacity helium vessel enclosed within a
radiation shield attached to a liquid nitrogen dewar. With the windows blocked
off and no electrical power input, the boil off rate is about 300 ml./hr.
With chopper running and windows open at float altitude, boil off is expected
to approach 1 l./hour. Since about half the helium is lost during ascent, this
allows 15 hours of flight time before the liquid level drops below the
detector.

2.3 Windows and Spurious Radiation

The radiation enters the cryostat through two windows as shown in Fig.1.   The

innermost window is Melinex and is 0.25 mm thick. Its transmission spectrum
is well known and contributes to the overall system response function
(c.f. section 2.5).   Since it is in contact with liquid helium, its
emissivity is unimportant.

However, in order to prevent the residual atmosphere from freezing onto the
cold window, another window is required at ambient temperature (~ 250 K).
This outer window is polyethylene film, 25μm thick and because of its high
temperature its emissivity is very important.   The latter has been deduced
from transmission measurements on thick samples of polyethylene.   These
results agree well with those reported by Chantry et al.[3]. The expected
window emission is shown in Fig. 2 along with a 2.7 K Planckian spectrum
for comparison.

A double window system introduces the possibility that spurious radiation from
large angles may enter the instrument acceptance cone due to multiple
reflections between the surfaces.   The design of the window system including
an absorbing mask which surrounds the helium window ensures that for each pair
of reflections this spurious radiation is attenuated by a factor of ~ 100.
The surfaces have been brought close together to increase the multiplicity of
the pairs of reflections.

Because the thin polyethylene window is unable to withstand a large pressure
differential, it must be covered by a protective vacuum sealed plate which is
only opened at float altitude. Attached to the cover is a polished copper
shield which reduces the radiation incident into the helium window/mask due to
earthshine. This lowers the helium boil-off rate as well as the detectable
spurious radiation content. Angular response measurements with an external
source indicate that spurious radiation should contribute less than that due
to window emission.

2.4   The Detection System

The detecting element is an InSb (hot electron) bolometer operating in the
Rollin mode, and is in the form of a slice $25 \times 5 \times 0.7$ mm^3 with two end
contacts. The contacts are formed with sulphur doped indium using the
radiation wetting process[5].

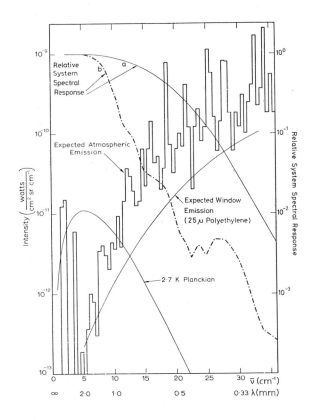

Fig. 2. The intensity scale refers to the 2.7 K Planckian curve, the expected window emission and the expected atmospheric emission. The system spectral response is normalized to 1.0 at $\tilde{\nu}$ = 5 cm^{-1}. Curves (a) and (b) respectively refer to the response without and with the mesh filter inserted into the beam.

A constant bias current is applied to the detector and the output voltage from the elements is coupled to the input of a cooled transformer which noise matches the detector to the external electronics. The transformer primary is series tuned by the dc blocking capacitor and the secondary is parallel tuned to the chopping frequency of 126.7 Hz. The transformer, dc blocking capacitor and the constant current resistor are mounted close to the detector and enclosed in a superconducting lead box which is overlaid by a further screen on mumetal. The output from the transformer is fed via a tight-twisted mumetal screened pair to the top of the cryostat where the low noise preamplifier is situated. The preamplified signal is then passed to the lock-in amplifier system.

The noise contribution by each element of the detection system has been
determined with the detector and the cold electronics at 1.8 K. The
following noise data referred to the transformer input were obtained:-

a)	Detector	0.12 ± 0.02	nV Hz$^{-\frac{1}{2}}$
b)	Transformer	0.25 ± 0.01	nV Hz$^{-\frac{1}{2}}$
c)	Preamplifier and lock-in	0.10 ± 0.01	nV Hz$^{-\frac{1}{2}}$
d)	Overall noise	0.30 ± 0.02	nV Hz$^{-\frac{1}{2}}$

Although the noise is dominated by the cooled transformer, omission of this
component would raise the contribution of the preamplifier and lock-in to
3 nV Hz$^{-\frac{1}{2}}$.

2.5 The Chopper

The chopper is a Perspex disc, 15 cm diameter and 1 mm thick upon which 19
segments have been aluminized. The disc is mounted directly on the drive
shaft of a cryogenic motor. This motor consists of a conventional low
voltage hysteresis synchronous motor in which the original phosphor bronze
sleeve bearings were replaced by miniature stainless steel ball races. The
motor is wrapped in lead foil to reduce leakage of the AC magnetic field.

Conventional hysteresis synchronous motors are self starting provided the
motor is able to accelerate to the synchronous speed within five cycles of the
applied drive frequency. The disc is designed to rotate at 400 rpm using a
drive frequency of 80 Hz but the inertia of the chopper disc stalls the motor
if the motor is started from standstill using an 80 Hz supply.

Therefore a special drive circuit has been developed which starts the motor
by applying a 10 Hz drive. When the motor is in synchronization the frequency
of the drive is increased slowly from 10 to 80 Hz. At this point, the 80 Hz
drive is then locked to a submultiple of the telemetry clock.

The acoustic noise generated by the chopper bearings is chiefly at high
frequency. The detected electrical noise in a bandwidth of 2.5 Hz centered
about the chopping frequency of 126.7 Hz is very small and increases the

detector system noise from 0.30 to 0.34 nV Hz$^{-\frac{1}{2}}$.   The helium loss due to the chopper in motion is 285 ml per hour (i.e. heat input ~ 200 mW).

2.6 Calibration and System Performance

Primary calibration is achieved by replacing the protective cover shown in Fig.1 by a subsidiary cryostat containing a black body cone which is orientated along the optic axis. The cone is made of Eccosorb 110, an iron-loaded resin, whose optical properties (at room temperature) have been measured. For optically thick samples, the Eccosorb 110 emissivity ranges from 0.8 - 0.9 depending on the sample density. The surface of the cone is specularly smooth so that most radiation emitted onto the cone from the hot window frame is absorbed after a few reflections rather than being scattered back into the acceptance solid angle.   Emissivity of the cone based upon the calculation of Lin and Sparrow[5] is greater than 0.95. The cone is thermally clamped to a copper backing cone which terminates in the helium dewar of the subsidiary cryostat. Five calibrated Allan-Bradley resistors are embedded in the Eccosorb in order to monitor temperature uniformity.

With liquid N_2 in the helium dewar calibration is successfully achieved with a black body temperature of 77 K. At helium temperatures, however, satisfactory temperature uniformity has not yet been obtained. Thus a secondary calibration method has been devised incorporating an Eccosorb 110 disc placed on a support which is clamped to the helium window frame of the main cryostat. The temperature of the disc can be varied between 2.5 K and 13 K. With an assumed emissivity of 0.85, these low temperature measurements are compatible with those obtained with the black body of 77 K.

The relative spectral response of the system is shown in Fig.2, curve (a). It is the product of all the spectral contributions along the optical path. The rapid fall of response above $\tilde{\nu}$ = 20 cm^{-1} is due mainly to the detector, a Fluorogold filter (thickness $\simeq$ 2.3 mm) placed in front of the detector, and the Perspex chopper (thickness = 1 mm). The relative response curve is normalized to 100% at $\tilde{\nu}$ = 5 cm^{-1}. At this point the absolute response (referred to the detector output) is ~ 8 V cm^2 sr W^{-1}. Together with the system noise $\sim 3.10^{-10}$V/ (Hz)$^{\frac{1}{2}}$, this leads to a signal/noise ratio of 10:1 at the peak of the 2.7 K curve for a 5 minute run time and a resolution interval

$\Delta \tilde{\nu} = 1.0 \text{ cm}^{-1}$.

In order to avoid detection of excess radiation at frequencies above $\tilde{\nu} = 10 \text{ cm}^{-1}$ (e.g. from the residual atmosphere) which might saturate the ADC, thus masking the 2.7 K background radiation, it is possible to insert a mesh filter into the beam. This is located in a rotatable filter wheel in front of the chopper (Fig.1). The system response with this filter included is shown in Fig.2, curve (b). The filter has a relatively flat pass band region for $\tilde{\nu} < 8 \text{ cm}^{-1}$ and then cuts off sharply above $\tilde{\nu} = 9 \text{ cm}^{-1}$. However, at helium temperatures the rejection by the multi-element filters changes significantly, presumably because of anisotropic contraction of the films. Introduction of the filter reduces the absolute response by about 20%.

The effectiveness of the filter in removing the expected atmospheric emission is shown in Fig.3.

3.   Emission from the Residual Atmosphere

The atmosphere emits because there are sources of opacity (spectral lines) which emit radiation in keeping with Kirchoff's law. The emission makes direct measurement of the sub-millimetre background around and above the frequency corresponding to the peak of 2.7 K black body spectrum impossible from the ground and difficult from balloon altitudes. This is in spite of the fact that balloons can carry equipment to heights which are above 99% of the atmosphere.

At these altitudes, 99% of the residual atmosphere above the balloon is within a vertical distance of 35 km. Therefore it is the atmospheric radiative processes in this region that are of great importance for the cosmic background measurements.

The molecular species which contribute significantly to the spectrum are H_2O, O_3 and O_2 since only these have sufficiently large column densities and lines of sufficient strength. For H_2O, the volume mixing ratio in the upper stratosphere is around 4×10^{-6} perhaps slowly increasing with altitude[6]. For O_3 the mixing ratio falls off rapidly in the upper stratosphere. The

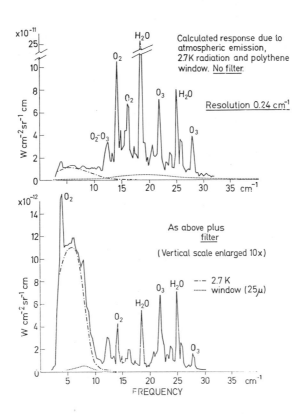

Fig. 3. Calculation of the expected spectrum due to the residual atmosphere, 2.7 K Planckian background and outer window. Individual contributions of the latter two are shown as dotted and dashed lines. Unapodized resolution is $\Delta \tilde{\nu} = 0.24 \, \text{cm}^{-1}$. A sinc^2 resolution function was used in the convolution.

abundance is not accurately known and may be variable[7]. Finally for O_2 the mixing ratio of molecular oxygen is constant[8] at 0.21 to at least 80 km. The vertical column density can therefore be determined from the pressure at the point of observation.

3.1 The Atmospheric Model

The residual atmosphere is approximated by an isothermal, horizontally stratified gas with pressure decreasing exponentially with altitude. It is taken to be isothermal because the absolute temperature of the atmosphere varies only by 10% in the region of interest above the balloon[9]. Since observations will be restricted to angles of viewing well above the horizon and the distance to the atmospheric emitting molecules is small compared with radius of the earth, horizontal stratification is an allowable approximation.

The emission lines are pressure broadened. Doppler broadening is insignificant in the spectral range of interest for pressure ≥ 0.1 mb (i.e. important for only the few percent of the residual atmosphere above ~ 4 mb). An altitude dependent Lorentz lineshape is used, the physical line widths being $\sim$ few $\times 10^{-4}$ cm^{-1}.

The equivalent width of the spectral line under the above conditions for species of constant mixing ratio is

$$\Delta \tilde{\nu} \simeq S \overline{N} \left(1 + \frac{S\overline{N}}{2\overline{p}\delta_o} \right)^{-\frac{1}{2}}$$

where $\overline{N}$ is the column density (molecule cm^{-2}), S the line strength (cm^{-1} molecule^{-1} cm^2), $\overline{p}$ the pressure at the observation altitude and δ_o is the Lorentz half width at standard pressure. This approximation is accurate to 2%.

The intensity in the emission line is given by

$$I(\tilde{\nu}) = b(\tilde{\nu}, T) \Delta \tilde{\nu}$$

where $b(\tilde{\nu}, T)$ is the Planck function for the atmospheric temperature T at the line frequency $\tilde{\nu}$.

Although ozone is not constantly mixed, lines are expected to be unsaturated and therefore the equivalent widths are independent of its distribution (and also lineshape).

Knowing the frequencies, strengths, and halfwidths of the lines [10,11] and the column densities of the species above the observation level, the expected atmospheric emission intensity has been calculated for the following parameters:

$$\overline{p} \; = \; 4 \; \text{mb, zenith angle} = 30^{\circ}, \; \overline{N}_{H_2O} = 4.9 \times 10^{17} \; \text{molecule cm}^{-2},$$

$$\overline{N}_{O_3} = 3.8 \times 10^{17} \; \text{molecule cm}^{-2}, \; \overline{N}_{O_2} = 2.1 \times 10^{22} \; \text{molecule cm}^{-2}.$$

The lines have been integrated into 0.5 cm^{-1} bins and are plotted as a histogram in Fig.2.

The spectra which should therefore be observed are obtained upon multiplying the emission spectrum by the response of the optical system (curves (a) and (b) of Fig. 2) and convolving with the instrument resolution function. In Fig. 3 we show apodised spectra (unapodised resolution 0.24 cm^{-1}) with and without the filter inserted into the beam. The large contribution of the residual atmosphere is obvious and indicates that detailed correction for atmospheric emission is of prime importance in any experiment of this type.

4. Stabilized Balloon Platform

The cryostat is mounted in a stabilized platform similar to that designed and flown by University College London. The platform provides a strong rigid framework in which to mount the cryostat so that it can be rotated in the vertical and azimuth planes. An outer protective framework contains the electronics and batteries. Azimuth control is obtained by a D.C. torque motor mounted in the vertical axis above the platform frame, which rotates the platform against an inertial wheel. Elevation angle control of the cryostat is obtained by another torque motor mounted on the horizontal axis bearing shaft.

The azimuthal angle is determined with respect to the earth's magnetic field by using a magnetometer, and the elevation angle by a precision potentiometer geared to the torque motor shaft. The overall accuracy of control is better than 1° both in elevation and azimuth.

The platform weighs ~220 Kg and is $1.7 \times 1.7 \times 3$ m high. The total weight of the platform and experiment is ~500 Kg.

The experiment is due to be flown from Palestine, Texas in August 1975.

Acknowledgments

The support of the Science Research Council is appreciated and one of us
(Paul Marchant) acknowledges the award of a Science Research Council Research
Studentship.

References

1. Robson, E.l., Vickers, D.G., Huizinga, J.S., Beckman, J.E. and Clegg, P.E.,
    1974, Nature, 251, 591.
2. Woody, D.P., Mather, J.C., Nishioka, N.S., and Richards, P.L.,   1975,
    Phys. Rev.Lett. 34, 1036.
3. Chantry, G.W., Fleming, J.W., Smith, P.M., Cudby, M. and Willis, H.A.,
    1971, Chem.Phys.Lett. 10, 473.
4. Lin, S.H., and Sparrow, E.M., 1965, J. Heat Transfer, 299.
5. Barber, H.D., Heasell, E.L., 1965, Solid State Electron. 8, 1130.
6. Martell, E.A., 1973, Physics and Chemistry of the Upper Atmosphere,
    McCormac, B.M. (Ed.), 24-33 (D.Reidel).
7. Carver, J.H., Horton, B.H., O'Brien, R.S., Rofe, B., 1972, Plan.Space
    Sci. 20, 217-223.
8. Handbook of Geophysics and Space Environments, 1965, Shea L. Valley (Ed)
    (McGraw-Hill).
9. U.S. Standard Atmosphere Supplements 1966 (U.S.G.P.O. Washington D.C.)
10. McClatchey R.A. et al., 1973, A.F.C.R.L. Atmospheric Line Parameter
    Compilation, A.F.C.R.L. -TR-73-0096.
11. Gebbie, H.A., Burroughs, W.J. and Bird, G.R., 1969, Proc.Roy.Soc.A. 310,
    579-590.

DISCUSSION

Traub Did you include stimulated emission in the line strengths, especially
in so far as your temperature differs from the AFCRL value?
Marchant The full expression for the temperature dependence of the line
strengths has been used which includes the factor $(1 - \exp - h\nu/kT)$. This
factor gives rise to an inverse linear temperature dependence for the line
strength in the spectral range of interest ($h\nu \ll kT$).
Traub What is the effect on emission strength of a more realistic non-
isothermal atmosphere?
Marchant No deviations from an isothermal atmospheric model have as yet
been considered, although the matter will be investigated. However, since
the emitted intensity of an isothermal atmosphere is not strongly dependent
on temperature, one would expect only a small change.

THE COSMIC BACKGROUND SPECTRUM
BETWEEN 0.7 AND 3.0 MM

E. I. Robson

Queen Mary College, London E.1

Abstract   A liquid cooled twin-beam polarizing interferometer
was flown by balloon to an altitude of 40 km on March 13th 1974.
The instrument was sensitive to radiation in the wavelength
region 3 mm to 350 μm. Interferograms were obtained which when
Fourier transformed appear to show the turnover of the cosmic
background radiation at wavelengths shorter than 1 mm and
correspond to a blackbody temperature of around 2.9 K.

To measure the critical region around the suspected turnover of the 2.7 K
cosmic background radiation, more sophisticated instrumentation than multi-
filtered photometers is required. This leads one into the realms of
interferometry and flown, in the first instance, from a balloon platform to
give a few hours of observations hopefully above the bulk of the atmosphere.

Design criteria for maximum energy throughput, and an apodised resolution of
0.5 cm^{-1}, constrained by the need to cool the entire interferometer to liquid
helium temperatures, led to the optical arrangement shown in Fig. 1.
Incident sky radiation enters the instrument via a thin, low emissivity
window in the cryostat vacuum jacket Fig 2. The first stage of the
polarizing Michelson interferometer[1] is a chopper which polarizes and
modulates the radiation. The beam is deflected into the instrument by a
plane mirror and is collimated by a spherical mirror at 45°. The beamsplitter
comprises a second polarizing grid, of 100 μm grating interval (the same for
all grids which are wire wound freestanding systems[2]. Two perpendicularly
polarized components of the beam are separated, the component with its E
vector perpendicular to the wires being transmitted and the parallel
component reflected.   These components are each reflected from the roof
mirrors which rotate their planes of polarization through π/2. They are then
respectively reflected and transmitted by the beamsplitter, recombining to

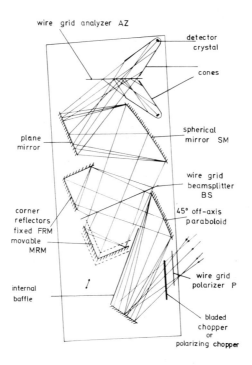

wire grid analyzer AZ

detector crystal

cones

plane mirror

spherical mirror SM

wire grid beamsplitter BS

corner reflectors fixed FRM movable MRM

45° off-axis paraboloid

internal baffle

wire grid polarizer P

bladed chopper or polarizing chopper

<u>Fig. 1</u> Optical arrangement of polarising Michelson interferometer. When a polarising chopper is used, the polariser P is omitted.

form a single output beam.   Interferometric scanning is performed by moving the lower pair of roof mirrors through 1 cm each side of the zero path difference position. The output beam is directed to a wire grid analyzer, which splits the beam into two interferometric components, sampled by two indium antimonide Rollin detectors at the apexes of the conical optics. With this arrangement, not only are both output ports used[3], but we have the safeguard of having two detectors; should one fail, only half the observing time is lost. The second input port is located internally in the interferometer and a blackened copper disc is mounted in this beam. The disc is mounted to a copper baffle plate which is thermally anchored to the helium radiation shield of the cryostat. A heater and thermocouple had been attached to the rear face of this disc to enable it to perform as an in flight calibration. Numerous tests, however, showed that this device was not

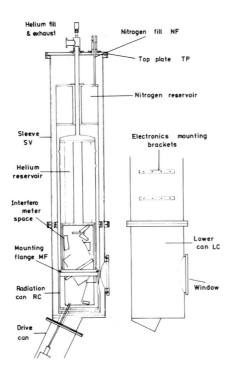

Helium fill
& exhaust

Nitrogen fill NF

Top plate TP

Nitrogen reservoir

Sleeve
SV

Electronics mounting
brackets

Helium
reservoir

Interfero
meter
space

Mounting
flange MF

Lower
can LC

Radiation
can RC

Window

Drive
can

Fig. 2   Interferometer mounted in the liquid helium flight
cryostat.

sufficiently accurate to act as a useful calibrator and so for flight, the
heater and thermocouple were removed. The transformer for the Rollin
detectors are mounted in magnetically shielded cans on an enclosing top plate
of the interferometer and as with all temperature-critical regions, including
the detectors, are thermally strapped to the liquid helium bath with thick
copper braids. A field stop, mounted on the interferometer side of the
polarizing chopper, limits the cone angle of the beam to 3°, giving, with
the collimating mirrors, an area-solid angle product of 10^{-5} m^2 sr.

In operation, the interferometer lies in a vertical plane and operates in a
"dry" environment, cooling being achieved by thermal conduction rather than
immersion in liquid helium. Interferometer components are temperature

monitored and with the addition of the aforementioned thermal braid heat
sinking, all were found to be attaining temperatures within a degree of the
liquid helium bath temperature. The cryostat contains 17 litres of liquid
helium, and 8 litres of liquid nitrogen for further shielding.

Three areas of operation are decidedly tricky and deserve a mention. First,
the chopper, originally an occulting blade system mounted behind a polarizer,
but replaced with a wire wound polarizing chopper[3]. There were three main
problems. Driving the chopper shaft meant having a motor which could not be
mounted within the cryostat and so a magnetic coupling was taken through the
aluminium wall of the cryostat. This posed difficulties because of eddy
currents and jitter when the chopper was being driven up in incremental
frequency steps to its operating shaft speed of 45 Hz. Also the reference
signal for the P.S.D. must be obtained from the chopper and fibre optics were
used to monitor this.

The second problem was the interferometer drive system. The motor, which
operates at 350 K, is mounted in a lower, uncooled vacuum can and thermally
isolated from the cryogenic regions of the dewar. The drive shaft comprises
point contact universal drive mechanism plus twin tufnol insulators and a
multiconvoluted stainless steel shaft. This drives through a micrometer screw
to a point contact on the interferometer carriage. Finally, because of the
internal vacuum of the cryostat, one has the thin fragile radiation input
window, which must be protected from pressure differences greater than 20 torr.
This necessitates a complex system of valves and explosive devices to ensure
it is protected until float altitude is attained when the protecting cover
MUST be removed by an automatic sequence. The above problems are more
fully described elsewhere[4].

Calibration is performed by mounting a variable temperature blackbody over
the entrance aperture of the cryostat as shown in Fig. 3. The blackbody is
an internally treated cone calibrated by a method suggested by the Cornell
group[5] and may be held at room, liquid nitrogen, liquid hydrogen and
various liquid helium temperatures. This enabled the response function of the
interferometer to be obtained and to understand the exact contribution of the
second input port, a point which must be considered in interferometer
applications.

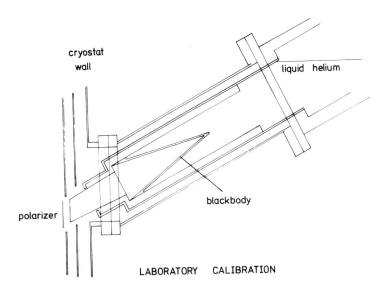

cryostat
wall

liquid helium

blackbody

polarizer

LABORATORY   CALIBRATION

Fig. 3   Calibration cryostat mounted in place of vacuum window
of the flight cryostat.

The first launch of the experiment occurred from Palestine, Texas, in May
1973.   This was abruptly terminated at an altitude of 60,000 ft when the
balloon burst and the payload suffered a somewhat retarded free fall to land
in a forest.   Disaster was averted by the cryostat's safety vent valves and
because by good fortune the dense tall trees caught the balloon rigging
wires, unopened parachute chords and balloon cable and prevented the payload
burying itself in the ground.   It was found hanging six feet from the ground!

The second flight was from the same launch site on March 13th 1974.   A float
altitude of 39.6 km was attained corresponding to an atmospheric pressure of
2.8 mb.   A spectrum with the window cover plate in position was recorded
during ascent and is shown in Fig. 4 illustrating that the system was working
well.   The window cover removal sequence operated successfully and 30
interferograms were obtained at the full resolution of the instrument plus a
further 10 at reduced resolution.   Unfortunately, because of a telemetry
malfunction, 60% of the time at float altitude was lost to data collecting.
The results of the flight are presented in ref (6);   an average of 10
unapodised 0.25 cm^{-1} resolution spectra is shown in Fig. 5.   We believe this
clearly shows the turnover of the cosmic background spectrum in the region

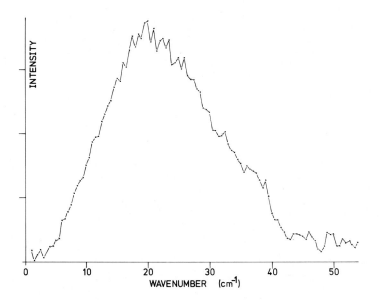

Fig. 4 Spectrum obtained with window cover in place during ascent.

between 10 cm^{-1} and 12 cm^{-1} before the atmosphere and window emission becomes
dominant. We obtain a temperature for the cosmic background radiation of
2.9 K as shown in Fig. 6.   Since the QMC flight, the Berkeley group have
also flown a polarizing Michelson interferometer and recorded spectra in the
same wavelength range. Although their conclusions are similar to those of
QMC regarding the temperature of the cosmic background radiation (Woody et al[7]
obtain a temperature of 2.99 $\pm$ $^{0.07}_{0.14}$) the spectra appear quite different.
Berkeley do not observe the turnover dip beyond 10 cm^{-1} but a radiation
continuum extending to higher frequencies. The performance of the apparatus
of the groups are very similar, differences being that Berkeley had an
immersed system with no window. Fig. 7 shows a spectrum of the background
plus atmosphere deduced for the parameters used by Woody et al. but at a
temperature of 250 K, plus window, superimposed on the result obtained by
QMC.   Flights were in the winter for QMC and summer for Berkeley (July 24th
1975) and at different altitudes,2.7 mb to 3.0 mb QMC and 3.2 mb to 3.4 mb
for Berkeley.

In conclusion, the present status is that both groups claim a cosmic
background temperature of around 3 K but disagree upon the interpretation of
atmospheric parameters in obtaining this figure.

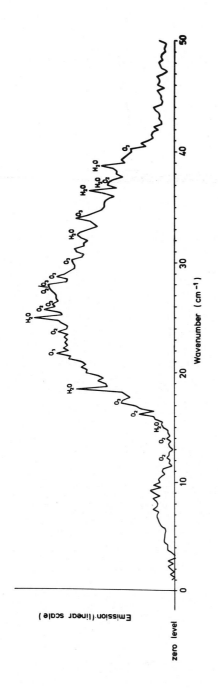

Fig. 5 Average of 10 raw spectra, uncorrected for instrumental response function, at 0.25 cm⁻¹ resolution (unapodized)

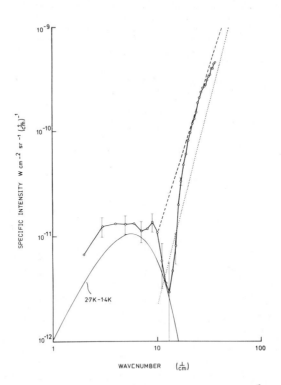

<u>Fig. 6</u> Cosmic background spectrum at 1 cm^{-1} resolution; –O–O–
observed spectrum; – – – – predicted window emission; . . . .
window emission using the results of Chantry et al. (1971).
Smooth solid line denotes 2.7 K – 1.4 K blackbody curve.

References

1. Martin, D.H., Puplett, E., 1971, <u>Infrared Physics</u>, 11, 1.
2. Vickers, D.G., Robson, E.I., Beckman, J.E., 1972, <u>App.Optics</u>, 10, 682.
3. Martin, D.H., 1972, <u>Infrared Detection Techniques for Space Research</u>,
                         eds. V. Manno and J. Ring (D.Reidel, Holland).
4. Robson, E.I., 1973, Ph.D. Thesis, University of London.
5. Shivanandan, K., Houck, J.R., Harwit, M., 1968, <u>Phys. Rev. Lett.</u>,
                                                     <u>21, 1460.</u>
6. Robson, E.I., Vickers, D.G., Huizinga, J.S., Beckman, J.E., Clegg, P.E.,
                         1974, <u>Nature</u>, 251, 591.
7. Woody, D.P., Mather, J.C., Nishioka, N.S., Richards, P.L., 1975,
                         <u>Phys.Rev.Lett.</u>, 34, 1036.

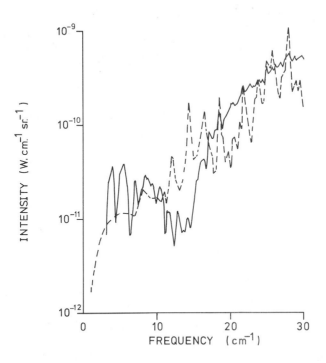

Fig. 7 Observed (———) and predicted (– – – –) spectra for the
QMC flight.   The predicted spectrum is based on the column
densities of Woody et al.(1975) and the window emission
measurements of QMC. Oxygen lines below $10cm^{-1}$ have been omitted.
The observed spectrum has been fitted to the model at the peak of
the $21.8cm^{-1}$ O_3 feature.

DISCUSSION

Coron Was the secondary input of the interferometer the same in the two
spectra you compared?
Robson Yes.
Edmunds What limits can you put on your measured blackbody temperature?
Robson 2.7 K plus 0.1 K, minus 0.3 K.
Edmunds Is this assuming that it is a perfect blackbody spectrum?
Robson Yes.
Shivanandan Both your results and Richard's show a turnover at $10 cm^{-1}$;
however there is an increase at around $12 cm^{-1}$. Does this imply that ozone
emission in the stratosphere limits further submillimetre background
measurements from balloon experiments?
Robson Quite probably unless one can obtain very high resolution
interferometers and look between the ozone lines. These are very sharp and
with low resolution merely merge into a continuum.
Fazio Did you measure any anisotropy during the azimuth scans?
Robson This was not really possible as the payload rotated freely with the

balloon, but we saw no significant difference between interferograms.
Glass Are you interested in the direction you point at, other than up?
Robson We have no control over the rotation of the platform and the zenith
angle of the interferometer is fixed before launch.   We did ensure that
there was no way we could intercept the moon at any time, but as it
happened, launch was delayed and so the moon was not up.

MEASUREMENTS OF THE POLARIZED SKY BACKGROUND IN THE FAR INFRARED

A. Coletti

Istituto di Fisica dell'Atmosfera-CNR, Roma, Italy

and

F. Melchiorri and V. Natale

Laboratorio TeSRE-CNR, Infrared Section, Firenze, Italy

Abstract Observations carried out with a balloon-borne polarimeter
have shown an upper limit of about 1% for the polarization of the
Cosmic Background radiation in the 500-1500 μm wavelength region.
Moreover, some particular features of the atmospheric emission
are discussed.

1. Experimental Layout and Calibrations

A sketch of the balloon-borne polarimeter is shown in Fig. 1. It has been
described in detail in refs (1), (2). It consists essentially of the
following parts:

(a) A 20 cm reflecting telescope, f/4

(b) A rotating analyzer which modulates only the polarized component of the
infrared radiation in the 300-3000 μm region.

(c) A Germanium bolometer cooled down to about 2 K at the pressure of the
floating altitude.

(d) Two lock-in amplifiers 90° apart in order to measure both the intensity
and the plane of the polarized radiation.

(e) A magnetically oriented platform pointing S-N direction within 0.5
degrees. The signal output S can be written as

$$S = S_s + S_a p + S_p$$

where S_s is the polarized sky background (if any), S_a is the atmospheric
emission which is spuriously polarized by an amount p from the instrument

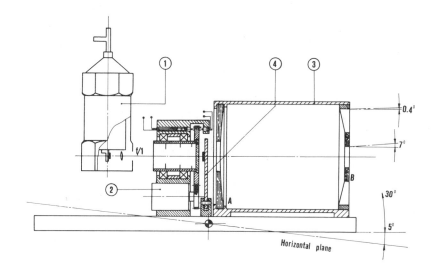

<u>Fig. 1</u>. Balloon-borne polarimeter: I-Ge Bolometer; 2-rotating
analyzer; 3-telescope; 4-on-board calibration source

S_p is the polarized emission from the instrument itself. Both p and S_p
have been measured in the laboratory by means of a reference source at
different temperatures. We have found $p = 5 \times 10^{-4}$ and
$S_p = (1.3 \mp 0.1) \times 10^{-10} T$ volt where T is the temperature of the analyzer.

2. Flight and Experimental Results

The polarimeter has flown from Air su Adour on September 16th, 1971 at 16 TU.
The floating altitude was reached at 18 TU. The signal v/s the time is shown
in Figure 2 for a fixed elevation of 25° above the horizon. We note the
following relevant features:

(1) The signal decreases and changes its sign at 17 TU. This fact indicates
that the atmospheric contribution decreases and finally becomes less important
than the instrumental background.

(2) A strong variation occurs between 17 and 17.30 TU; it is related to the
Sun which is in sight of the detector at this time; these data have been
considered in a previous work in evaluating the polarized Sun emission.(1,2).
The data between 18 and 20 TU allow us to evaluate the polarized sky background. We have measured in laboratory the instrumental background, but we
must correct the data from the residual contribute of the atmosphere. It has

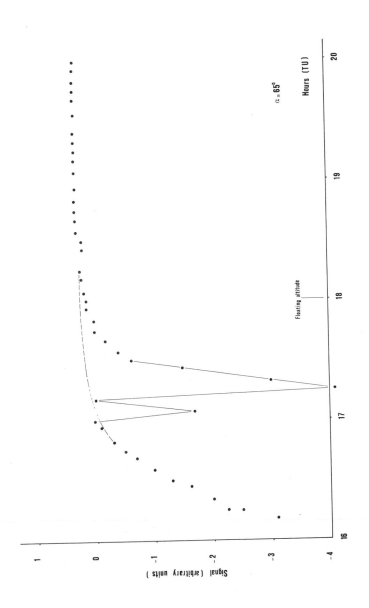

Fig. 2 Behaviour of the signal versus the time. The data have been corrected for the variation of the detector sensitivity with the altitude. The standard deviation for each point is of the order of 10%.

been accounted for by tilting continuously the instrument between sec ϕ= 2
and 12, as shown in Fig. 3. So the data of Fig. 2 corrected by subtracting
the instrumental background and the atmospheric emission given has an upper
limit which is zero at one standard deviation and 0.02 at two standard
deviations. Let us discuss briefly the unexpected knee at about sec ϕ= 7.
It cannot correspond to an effective absorption law since it will require
a transmittance of about 87%, while at the floating altitude of 30 Km the
transmittance must exceed 99%. We have considered two possible effects:

a. The signal is due to stray light coming from the earth, instead of an
atmospheric emission and it saturates as the earth becomes completely visible
from the instrument. This interpretation is in contrast with the angular
response of the polarimeter as measured in the laboratory (Fig.4).

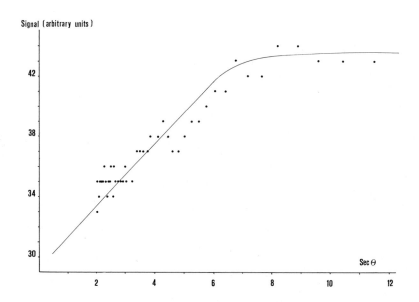

Fig. 3   A typical response of the polarimeter versus the secant
of the tilting angle; the knee at sec ϕ = 7 is discussed in the
text.

b. The signal is produced by a cloud of aerosol with particles at high
temperature (400 K or more) as suggested by (3). In this case the evaluated
extinction is of the order of 0.16, assuming a temperature of about 400 K for
the grains. However, the density required in order to obtain the knee is too
high (about 1pcm^{-3}).

We are carrying out a systematic study of the data in order to obtain more
information about the upper limit of the polarized sky background and on the
possible causes of the observed behaviour of the atmosphere.

References

1. Dall'Oglio,G; Melchiorri, B; Melchiorri, F; Natale, V; Gandolfi, E., 1973
       "Evidence for polarized emission from the Sun in the far infrared";
       Infrared Phys. 13, 1.
2. Dall'Oglio, G; Melchiorri, B; Melchiorri, F; Natale, V., 1974.
       "Astronomical polarimetry in the far infrared"; Planets, Stars and
       Nebulae studied with Photopolarimetry; Gehrels Ed; Tucson Arizona.
3. Grams, G; Fiocco, G; Mugnai, A., 1975, "Energy change and temperature of
       aerosols in the earth atmosphere" NCAR 1307-75-6.

DISCUSSION

Fazio Have you ever noticed any water vapour associated with the balloon?
Melchiorri Since we observed the sky at maximum elevation of 30 km above the
horizon, we were not able to detect emission from the balloon. However, in a
previous flight we observed a strong signal near the zenith which could be
due either to water vapour or to earth radiation reflected from the balloon.
Sollner Could the flattening in the sec θ plot you observed be explained by
the curvature of the atmosphere?
Melchiorri The curvature of the atmosphere can help, but one should note that
this knee appears at an elevation of 70°. I want to emphasize that this
behaviour of the atmospheric emission observed by us has not been seen by
Hoffmann, so that we are forced to interpret it as a local effect, i.e.
instrumental or due to a local atmospheric feature.
Marchant Do you have any further information on the aerosols you mentioned?
Melchiorri The expected dimensions of the particles are between 0.01 and
1 μm and the density at an altitude of 30 km has to be less than 0.1 cm^{-3}.
The temperature, evaluated as a balance of radiative and conductive effects,
is of the order of 450 K. In order to evaluate the emission, we have
adopted the complex refractive index given by Ivlev and Popova. However,
this refers to lower troposphere aerosols, so a large error may arise in our
computation from this assumption, since the absorption efficiency is directly
proportional to the imaginary part of the refractive index. Unfortunately
the refractive index cannot be evaluated from the laser observations which
have detected the aerosols in the upper atmosphere.
(ref: Ivlev L.S. and Popova S.I., 1973, Izv. Atmosph. and Oceanic Phys.
                          9, 1034-1043)

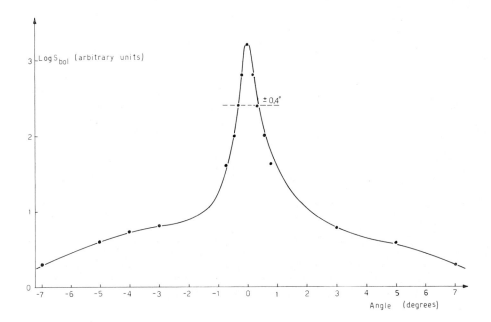

Fig. 4.   The angular response of the polarimeter as measured in the laboratory by means of a point-like source; the wings are due to the large field of view through the central hole in the secondary mirror, as shown in Fig. 1.

LIMITS ON A MICROWAVE BACKGROUND WITHOUT THE BIG BANG

J. V. Narlikar,[*] M. G. Edmunds and N. C. Wickramasinghe

Department of Applied Mathematics and Astronomy,

University College, Cardiff, Wales

Abstract The possibility of explaining the cosmic microwave background in terms of thermalisation of radiation from such sources
as galaxies by dust grains is explored further. Relevant calculations of the optical cross-sections of graphite whiskers are given
and it is shown that a smeared out dust density of $\sim 10^{-33}$ g cm^{-3}
is required. Limits are set on the large-angle anisotropy of the
background which is to be expected on the basis of this model.
The relative merits of the conventional explanation and the present
theory are discussed and a few discriminatory observational tests
proposed. Some cosmological implications of whisker grains in the
intergalactic space are examined.

1. Introduction

A universal background radiation existing as a relic from the big bang was
predicted by Gamow and his collaborators[1] in the 1940's, on the basis of their
work on nucleosynthesis in the very early stages of the universe. Subsequently
Dicke and his colleagues[2] at Princeton reaffirmed a similar prediction. Whilst
the thermal nature of the background was clearly expected in these predictions,
its temperature could be specified only crudely. The original measurement
of background radiation by Penzias and Wilson[3] indicated a temperature slightly
in excess of 3 K, although subsequent measurements by other observers at
different wavelengths set the best value of the temperature at 2.7 K. These
observations are to the long wavelength side of the expected black body peak.
Measurements at the peak of the curve are as yet tentative, and there are no
undisputed published observations at shorter wavelengths. Nevertheless, the
fit of the observed spectrum is sufficiently good to generate a confidence

[*]*On leave of absence from the Tata Institute of Fundamental Research, Bombay 400005, India.*

among most astronomers that the background has a blackbody distribution
peaking at the microwave wavelength of $\sim$ 1 mm.

The success of the big bang explanation has tended to overshadow other
possible candidates for this background. Attempts to account for the
background in terms of sources[4,5] or other processes taking place at more
recent epochs[6] have encountered serious objections, particularly with regard
to the observed fine scale isotropy[7,8]. To avoid any patchiness the sources
of radiation have to be far more numerous than galaxies, although they need
not individually be exceptionally bright. An entirely new approach has been
suggested recently by Hoyle[9] in which he has related the microwave background
to the history of the universe prior to the big bang, by working in the framework of the Hoyle-Narlikar theory of gravitation[10].

In this paper we explore further the dust grain model proposed recently by
Wickramasinghe et. al[11]. In this model needle shaped grains such as the
graphite whiskers act as the thermalisers of radiation in the intergalactic
space. If the whiskers could grow to lengths in the range 100μm - 1 mm, they
would have large absorption cross sections for photons peaking over a range
of wavelengths $\lambda \gtrsim 0.3$ mm. Such grains could serve to thermalise the
incident radiation from galaxies, QSO's etc., and the resulting blackbody
curve could peak at a wavelength $\lambda \sim 1$ mm. This model gets round the
objections raised on the grounds of small scale anisotropy at mm wavelengths.
It may, however, be vulnerable to other observational or theoretical points
of criticism. We discuss some of these below.

2.   Thermalising Mechanism

The inadequacy of spherical or nearly spherical grains to serve as thermalisers
for the microwave background has already been discussed[12]. The main difficulty arises from the fact that the ratio of the absorption cross section at
optical wavelengths to that at infrared and millimetre wavelengths exceeds
unity by a large factor for grain radii $\leq \sim 10^{-4}$ cm. This is not necessarily
true for grains in the shape of long whiskers, which could occur in typical
galactic and intergalactic conditions.

We confine our discussion here to graphite whiskers. There is substantial documentation of whisker growth for this material (see Wickramasinghe et. al[11] and references therein) and we also have a fairly extensive set of optical data. Graphite whiskers may condense in the expansion phases following explosions of supermassive stars in galactic nuclei, as well as in the atmospheres of normal carbon stars. Their ejection into the intergalactic medium might be accomplished by the action of stellar radiation pressure, or by gravitational encounters of the type discussed earlier[11]. Alternatively, a pregalactic population of stars could provide carbon stars[13].

Consider graphite particles in the form of cylindrical needles of cross-sectional radius a and length $\ell \gg a$. The basal planes of graphite are parallel to the cylinder axis (Bacon[14]) and the complex refractive index data for electric vector along the cylinder axis are essentially those given by Taft and Phillipp[15]. The complex refractive index data for electric vector perpendicular to the basal planes are given by Tossatti and Bassani[16].

Cylinder cross sections are taken to have radii $a \simeq 10^{-6} - 10^{-5}$ cm, similar to the radii of interstellar grains. Whisker growth may allow particle lengths ℓ to become as long as several hundred microns (cf ref. 11). For optical and ultraviolet wavelengths we now have $2\pi a/\lambda \gtrsim 1$, and the rigorous formulae for cylinders must be used for computing optical cross sections at these wavelengths. The only condition necessary for the applicability of these formulae is $\lambda \ll \ell$, and this will be satisfied at optical wavelengths for cylinders of lengths $\gtrsim 1$ μm. For cylinders in random orientation, the mean efficiency factor for absorption, computed from the rigorous formulae for infinite cylinders, is

$$Q_{abs}(\lambda, a) = \frac{C_{abs}(\lambda)}{2\,a\,\ell} = \frac{1}{3}\left| Q_{abs}''(a, \lambda) + 2Q_{abs}^{\perp}(a, \lambda) \right| \tag{1}$$

Here Q'' refers to absorption efficiency calculated from the data of Taft and Phillipp[15] and $Q^{\perp}$ that from data of Tossatti and Bassani[16]. The values of $Q_{abs}(3200\ \overset{o}{A}, a)$ are given for various grain radii in the second column of Table I.

The asymptotic form of the mean absorption cross section in the limit of $2\pi a/\lambda \ll 1$ for a randomly oriented set of graphite cylinders is

$$C_{abs}^{IR}(\lambda) = -\frac{2\pi^2 a^2 \ell}{3\lambda} \, \text{Im}\{(m_{11}^2 - 1) + 4 \left| \frac{m_{\perp}^2 - 1}{m_{\perp}^2 + 1} \right| \} \,, \qquad (2)$$

provided $\lambda \stackrel{<}{\sim} \ell$. Writing $m_{11,\perp}^2 = K_{11,\perp} - 2i\sigma_{11,\perp}\lambda/c$ with the usual notation (K = dielectric constant, σ = optical conductivity) we obtain

$$C_{abs}^{IR}(\lambda > 300\mu) \qquad \frac{4\pi a^2 \ell \sigma_{11}}{3c} \qquad (3)$$

with $\sigma = 10^{15} \, s^{-1}$ (ref 11). For far infrared and millimetre wavelengths (3) is valid to a good approximation for graphite cylinders of radii $a \simeq 10^{-5}$ cm, lengths $\ell \sim 100\mu m - 1$ mm. The third column of Table I sets out the ratio $C_{abs}(\lambda > 300\mu)/C_{abs}(3200 \, \overset{o}{A})$ as a function of the cross-sectional radius a.

TABLE I

a/μ	$Q_{abs}(3200 \, \overset{o}{A}) = C_{abs}(3200 \, \overset{o}{A})/2a\ell$	$C_{abs}(\lambda > 300\mu)/C_{abs}(3200 \, \overset{o}{A})$
0.01	0.58	0.37
0.03	0.46	1.40
0.05	0.51	2.17
0.07	0.56	2.72
0.10	0.64	3.39
0.15	0.75	4.33
0.20	0.82	5.33
0.25	0.88	6.19

For grains of radii $> 0.10\mu m$ we note that the optical depth of the universe up to the Hubble radius at far infrared and millimetric wavelengths can be greater than that at ultraviolet wavelengths by a factor $\overset{>}{\sim} 3.4$. Thus the universe could be marginally optically thick in the UV but much more opaque at the far IR and millimetre wavelengths. This is therefore a feasible mechanism for thermalising optical and shorter wavelength radiation from galaxies, QSO's and other sources. On account of the multiple scatterings and reabsorptions of far IR photons an approach to an isotropic thermalised background is expected.

The mass density of whisker grains necessary to produce $\tau_{UV} \simeq 1$ at the Hubble radius $R \sim 2 \times 10^{28}$ cm can now be estimated. The mass absorption coefficient of whisker grains is

$$K(3200 \ \overset{o}{A}) = \frac{C_{abs}}{\pi a^2 \ell s} = \frac{2 \ Q_{abs}}{\pi \ as} \ cm^2 \ g^{-1} \tag{4}$$

For $s = 2g \ cm^{-3}$ and $a = 10^{-5}$ cm, equation (4) gives

$$K(3200 \ \overset{o}{A}) \overset{\sim}{=} 2 \times 10^4 \ cm^2 \ g^{-1} , \tag{5}$$

using Q_{abs} from Table I. The smeared out mass density of such grains required to produce $\tau(3200 \ \overset{o}{A}) = 1$ at the Hubble radius is therefore

$$\rho_{grains} = \frac{1}{KR} \overset{\sim}{=} 2.45 \times 10^{-33} \ g \ cm^{-3} \tag{6}$$

This is consistent with the limits on the intergalactic dust density which are determined from other criteria[17,18]. Also, this is two orders of magnitude lower than the smeared out density of matter in the form of galaxies, QSO's etc, and four orders of magnitude lower than the cosmological closure density.

3. Cosmological Considerations

Although this model is free at $\lambda \overset{<}{\sim} 3$ mm of the objections which can be raised on the grounds of small angle anisotropy[7,8] the question remains whether there could result an observable large angle anisotropy of the theramalised background. In particular, if the thermalisation is more effective for galaxies within a cluster (or a super-cluster) than for the field galaxies, would a nearby large cluster (or a super-cluster) produce a detectable fluctuation over large angles of the order of a few degrees? The expected fluctuation may be crudely estimated by the following simple calculation.

Suppose we are dealing with radiating units of linear size a and number density n. We shall later identify these units with clusters or super-clusters. Let each unit produce an energy output at the rate L per unit time, this being the energy available for thermalisation. To calculate the total background energy flux at a typical point it is necessary to specify the cosmological model. We shall consider two models: (a) the steady state model (SS in brief) and (b) the empty Friedmann model (EF in brief) to illustrate the results. The background flux density is given by

$$B = n \left(\frac{c}{H}\right) L \, f(z) \tag{7}$$

where c = velocity of light, H = Hubble's constant, and

$$f(z) = \begin{cases} \text{(a)} \quad \frac{1}{4}\{1 - \frac{1}{(1+z)^4}\} \quad \text{(SS)}, \\ \\ \text{(b)} \quad \frac{1}{2}\{1 - \frac{1}{(1+z)^2}\} \quad \text{(EF)} \end{cases} \tag{8}$$

For SS, $f(1) = 15/64$ while for EF $f(1) = 3/8$. If thermalisation takes place for $z \sim 1$, the two models do not differ from each other significantly.

Suppose we have a nearby unit at a redshift $z_o \ll 1$, contributing the radiation flux

$$b = \eta \frac{L}{4\pi(\frac{c}{H})^2 z_o^2} \qquad (9)$$

where η represents a fraction of the total radiation which has actually
thermalised. It must be remembered that as opposed to distant units the
light from a nearby unit may not completely thermalise (since thermalisation
involves long passage through the intergalactic medium). Thus the fluctuation
from the nearby unit is measured by the ratio

$$\frac{b}{B} = \frac{\eta}{4\pi(\frac{c}{H})^3 n f z_o^2} \qquad (10)$$

Notice that both η and f are less than unity in all cases. Although η
may be considerably smaller than f we will assume $\eta/f \sim 1$ and estimate an
upper limit to the fluctuation.

Suppose now that the units are clusters of $\sim 10^3$ galaxies and that on an
average the number of such clusters per $(Mpc)^3$ is $\sim 10^{-4}$. (This corresponds
to a smeared out average of 1 galaxy per $10 (Mpc)^3$. If an average galaxy
has a mass $\sim 10^{44} g$ the matter density in the form of galaxies works out at
$\sim 4.10^{-31} g \ cm^{-3}$.) With $c/H \simeq 6.10^3$ Mpc. we get

$$\frac{b}{B} \sim 5 \times 10^{-9} z_o^{-2} \qquad (11)$$

For the Virgo cluster we use $z_o \sim 4 \times 10^{-3}$ to obtain b/B as low as
$\sim 3 \times 10^{-4}$. For a super-cluster n is lower and z_o higher such that the
effect may be to raise b/B by an order of magnitude above this value. Such
a fluctuation is probably too small to be detected by presently available

techniques.

A futher question arises with regard to the ability of the units to cover the entire sky effectively. This is essential if all radiation is to be trapped and thermalised. For the units mentioned above the total solid angle subtended out to the redshift z is given by

$$\Omega \;=\; 4\pi^2 a^2 \left(\frac{c}{H}\right) n g(z) \tag{12}$$

where

$$g(z) \;=\; \begin{cases} \text{(a)} & \ln(1 + z) \quad \text{(SS)} \\ \text{(b)} & z\left(1 + \tfrac{1}{2}z\right) \quad \text{(EF)} \end{cases} \tag{13}$$

For clusters comprised of ~ 1000 galaxies, we may take $a \sim 5$ Mpc and get $\Omega \sim 180$ g. At $z = 1$, $g \sim .7$ for SS and ~ 1.5 for EF. Thus $\Omega \gg 4$ and the entire sky is expected to be covered. For superclusters Ω may be smaller but still much larger than 4π.

Finally, we come to the question of ascribing a temperature and spectrum to the thermalised radiation. If the spectrum were a blackbody spectrum the temperature is determined by equating aT^4 to the energy density of thermalised radiation (where a is the radiation constant). Using $L \sim 10^{47}$ erg s^{-1} and the other cluster parameters we get

$$aT^4 \;\sim\; 2 \times 10^{-13} \, f \; \text{erg cm}^{-3} \tag{14}$$

The expression on the right hand side is a little lower than the energy density for a blackbody of temperature $T = 2.7$, which is $\sim 4 \times 10^{-13}$ erg cm^{-3}. On the other hand it is somewhat higher than the energy density of starlight in the intergalactic space, which is usually quoted around a few times

10^{-14} erg cm^{-3}. This, however, is not a serious discrepancy. The mechanism discussed above uses not only starlight but other wavelengths of electromagnetic radiation like X-rays in thermalisation. The observations of extragalactic astronomy are still by no means complete in their discoveries of other sources of radiation than the optical emission from galaxies. The right hand side of (14) could therefore be an under-estimate.

Another check on this calculation is provided by the He/H ratio in the universe. If all the helium were synthesised in stellar interiors, the amount of starlight would be comparable to the microwave background[6]. As suggested by Wagoner, Fowler and Hoyle[19], the helium synthesis might well have to take place in supermassive stars.

4. Restrictions of the Theory

As thermalisation is taking place in an expanding space, the redshift effect will modify the shape of the thermalised spectrum. The spectrum observed at earth will be a superposition of contributions from different redshifts, and the spectrum will deviate somewhat from a single-temperature blackbody. We propose to investigate the spectrum in a future paper, but note here that the shape of the spectrum is likely to be a strong observational test of the theory.

Perhaps the most unsatisfactory aspect of the explanation is its inability to account for the background at wavelengths longer than a few millimetres as the absorption cross-section of the whiskers is likely to fall off as λ^{-2} for $\lambda > \ell$, giving only small optical depths out to the Hubble radius at radio frequencies. Three possible (although very ad hoc) modifications are:- (i) the whiskers grow to lengths > 10 cm. This would allow high optical depths at the typical wavelengths of isotropy measurements[20], but the fragility and growth times of such long whiskers makes their existence highly unlikely. (ii) A large enough density of dust and radiation at high redshifts so that a spectrum thermalised to $\sim$ mm blackbody spectrum has redshifted to provide the centimetre wavelength "tail" now. The present-day observed spectrum is again unlikely to be an exact blackbody. (iii) The conventional hot big-bang provides the centimetre tail, and the whisker thermalisation provides a contribution at wavelength $\leq$ 3 mm.

5. Conclusion

The big bang undoubtedly provides the simplest explanation of the cosmic microwave background. However, because of the profound implications of this result for cosmology, it is desirable, as a part of normal scientific practice, to consider alternatives to this explanation and discuss their relative merits. The present model has been proposed from this point of view.

The big bang explanation is not free from defects. For example, the present temperature (and energy density) of the radiation cannot be related to any other astrophysical property of the universe. The present temperature of 2.7 K and the coincidence of energy densities mentioned earlier must be regarded as accidental. An astrophysical explanation of the type considered here at least has the merit of relating the present temperature of the background to the processes taking place now or in the not too remote a past in the universe.

The second difficulty relates to the observed high degree of isotropy of the microwave background. The small particle horizons appropriate in the early stages after the big bang make it impossible to exchange information between remote parts of the universe. The present day isotropy cannot therefore be explained except by postulating it at the big bang. In the present approach the isotropy of the microwave background is related to the isotropy of the universe in the relatively recent past, irrespective of its isotropy (or lack of it) at very early epochs.

We realise that the grains required for our model are somewhat exotic and suffer from the constraints of a particular range of sizes. We have stated an extreme case in requiring the complete explanation of the microwave background by such grains, but an important consequence of these grains for the standard hot big bang models should also be noted. If the dust grain density were lowered by a factor $\sim$ 10 compared to that given in equation (6), the optical depths in millimetre and infrared would extend to redshifts $z \sim 2, 3$. This could still interfere with the isotropy of the original big bang radiation.

For example, the small scale fluctuations expected to lead to galaxy formation (at much earlier epochs than $z \sim 2, 3$) could be smoothed out by the intervening dust. We await with interest any results from observations attempting to detect such fluctuations at mm wavelengths.

Finally, we enumerate briefly some possible consequences of this model for extragalactic observations other than those of the microwave background. The absorption produced by the dust in the optical and UV may affect the measurement of the deceleration parameter (q_o) of the expansion of the universe. The result will be to give a q_o which is lower than the true value. The whisker grains will not have a high differential reddening effect in the optical or ultraviolet. If the QSO's are at the cosmological distances indicated by their redshifts their spectra are expected to show a serious modification in the millimetre range at large redshifts when compared to those at $z \ll 1$.

Acknowledgement

One of us (J.V.N.) is happy to acknowledge the grant of an S.R.C. Senior Visiting Fellowship at the Department of Applied Mathematics and Astronomy, University College, Cardiff.

References

1. Gamow, G., 1948, Phys.Rev., 74, 505.
2. Dicke, R.H., Peebles, P.J.E., Roll, P.G. and Wilkinson, D.T., 1965, Astrophys.J., 142, 414.
3. Penzias, A.A. and Wilson, R.W., 1965, Astrophys.J., 142, 419.
4. Narlikar, J.V. and Wickramasinghe, N.C., 1967, Nature, 216, 43.
5. Wolfe, A.M. and Burbidge, G.R., 1969, Astrophys.J., 156, 345.
6. Hoyle, F., Wickramasinghe, N.C. and Reddish, V.C., 1968, Nature, 218, 1124.
7. Conklin, E.K. and Bracewell, R.N., 1967, Nature, 216, 777.
8. Hazard, C. and Salpeter, E.E., 1969, Astrophys.J., 157, L87.
9. Hoyle, F., 1975, Astrophys.J., 196, 661.
10. Hoyle, F. and Narlikar, J.V., 1974, Action-at-a-Distance in Physics and Cosmology (San Francisco: Freeman).
11. Wickramasinghe, N.C., Edmunds, M.G., Chitre, S.M., Narlikar, J.V. and Ramadurai, S., 1975, Astrophys. and Space Sc., 35, L9.[*]
12. Longair, M.S. and Rees, M.J., 1973, in Cargese Lectures in Physics Volume 6, Ed. E. Schatzman, Gordon and Breach, London.
13. Layzer, D. and Hively, R., 1973, Astrophys.J., 179, 361.
14. Bacon, R., 1958, Cambridge Conference on Strength of Whiskers and Thin Films.
15. Taft, E.A. and Phillips, H.R., 1965, Phys.Rev., 138A, 197.

16. Tossatti, E. and Bassani, F., 1970, Nuovo Cimento, 65B, 161.
17. Nickerson, B.G. and Partridge, R.B., 1971, Astrophys.J., 169, 203.
18. Karachentsev, I.D. and Lipovetski, V.A., 1968, Astron.Zh., 45, 1148.
19. Wagoner, R.V., Fowler, W.A. and Hoyle, F., 1967, Astrophys.J., 148, 3.
20. Conkin, E.K. and Bracewell, R.N., 1967, Phys.Rev.Lett., 18, 614.

*Note important erratum in Astrophys.Space Sci., 35, No.2, 1975.

DISCUSSION

Greenberg Did you use a Rayleigh type approximation for the millimetre
emission from the long particles? If so did you take into account the length
limitation on the applicability of this approximation especially when m is
large. I tried to find the absorption efficiency magnification for metallic
particles and a factor of 10 or so was the most I could find.

Wickramasinghe Yes, we did use a Rayleigh approximation. Our results remain
valid for $\lambda > \ell/2$, say.

Greenberg The second point I would like to question is the use of 10^{-13} ergs
cm^{-3} for the visible-uv radiation density in intergalactic space. This
would be comparable with the visible-uv energy density in the Milky Way
and therefore rather on the high side, I think.

Narlikar The value usually quoted is a few times 10^{-14} erg cm^{-3} which is
lower than I have used. I feel, however, that there is sufficient uncertainty
in the measurements of extragalactic astronomy to permit the higher value.
Also it is consistent with the amount of starlight generated if all the
observed helium were generated in stellar nucleosynthesis, e.g. in Hoyle-Fowler
supermassive stars.

Hillas This model differs from the conventional one by filling space with
absorbers, and with an optical depth to the Hubble radius of 3 or 4 it
seems that the optical depth would be about 1 at redshifts of 0.25 say, so
one should expect to see a large bite out of the spectrum of quasars in the
millimetre region. I wonder what is the most distant quasar which has been
studied at millimetre wavelengths?

Narlikar Some such effect should show up in QSO's with large cosmological
redshifts (z $\gtrsim$ 0.5, say). I am not aware of any observational results of
this type at present.

GENERAL DISCUSSION ON INFRARED AND MILLIMETRE BACKGROUND

Chairman: Dr. K. Shivanandan

Chairman's Introduction

Near infrared background in the 10-30 μm band have been surveyed by Walker (Air Force Cambridge Research Laboratories) from rockets, Murcray from balloons and by an Air Force Satellite (Celestial Mapping Program). The results from these measurements are still not available in general to the infrared community. No measurements exist in the 30-1000 μm band. There is a microwave isotropic background radiation which has been interpreted as being thermal radiation from the evolution of the early universe. Monochromatic flux densities of this radiation measured at a number of wavelengths (λ = 3 mm to λ = 7 cm) indicate that this radiation is characterized by a 2.7 K blackbody spectrum, the peak of which is at 1 mm.

Measurements of the submillimetre background (λ < 1 mm) have been made using balloons, rockets and to some extent from ground based observatories. Early rocket flights found excess radiation in the 0.4 - 1.5 mm band (25-7 cm^{-1}), but later flights have reported results consistent with a 2.7 K blackbody spectrum. Broadband photometry using low pass filters in the 10-20 cm^{-1} region were carried out by an M.I.T. group using balloons at an altitude of 40 km. After subtraction of atmospheric background upper limits consistent with a 3 K blackbody radiation was obtained. Spectral measurements of the background using balloon-borne liquid helium cooled interferometers have been carried out by Queen Mary College, London (polarization interferometer: 3-40 cm^{-1}) and by the University of California, Berkeley (Michelson interferometer: 3-40 cm^{-1}). Both experiments indicate a turnover of the cosmic background at 10 cm^{-1} (λ = 1 mm) which seems to be consistent with a 2.7 K blackbody spectrum, but show excess radiation above 12 cm^{-1} (λ < 0.8 mm) due to atmospheric emission of ozone, oxygen and water. The University of Leeds groups is planning a balloon flight of a liquid helium cooled lamellar grating interferometer to measure the background in the 2-20 cm^{-1} region in August 1975.

In the context of present observations, it will be useful to analyse the results of the Queen Mary College and University of California observations and evaluate what can be achieved with the University of Leeds experiment. It is most unfortunate that there is no one from the University of California to present their results, however, I will call upon Queen Mary College group to outline their observational results and if possible compare it to the University of California published results and for the University of Leeds to define their planned experiment.

Summary of Q.M.C. Viewpoint (Dr. P. E. Clegg)

I shall first summarise the differences in the spectra observed by Berkeley and Q.M.C. and discuss possible reconciliation of the two results. Then I shall consider what we can say about the background should no reconciliation be possible.

Q.M.C. observe a large dip in the spectrum before correcting for the atmosphere, between 10 and 12 cm^{-1}, which they interpret as the turnover in the background spectrum. Berkeley on the other hand observe no such dip, simply a continuum which increases from the lowest observed frequencies. Secondly, Q.M.C. observe apparently less sharp atmospheric lines than Berkeley. In discussing these differences, the similarity between the two experiments should be borne in mind. Berkeley also used a Martin polarising interferometer, which indeed they adopted after inspection of the Q.M.C. design. The principal difference was that the Berkeley experiment operated in liquid helium while Q.M.C.'s was in vacuo, necessitating a vacuum window of very thin polyethelene. (It is believed that Berkeley originally had a window which ruptured on their first flight. Since this appeared to have no effect on their experiment, they discarded it for their second flight.)

In analysing their data, Q.M.C. fitted their expected window emission to the troughs of the atmospheric lines at the high frequency end. At lower frequencies, the window emission was less than expected, allowing observation of the dip in the spectrum. The main uncertainty in the Q.M.C. experiment was the window emission; it should not however affect the flux scale of the experiment by more than a factor of about 2.

Berkeley used a model spectrum with four adjustable parameters to fit their data. Their model atmosphere was isothermal with exponential density distribution and constant mixing ratio for the constituents H_2O, O_2 and O_3. The column densities of these, and the temperature of the background radiation were varied to obtain a "best" fit of the model with the observed spectrum, although no criterion is given for the quality of this fit.

How could Q.M.C. have obtained a dip when Berkeley observed none? A possibility is that the ozone concentration was different for the two experiments. (Berkeley were at a lower altitude (Robson, this volume, p.103) and had a summer flight, whereas we had a winter flight.) The ozone spectrum at 1 cm^{-1} resolution is relatively smooth and excess ozone could fill up the Q.M.C. dip. Why did Q.M.C. have apparently lower resolution? Apart from instrumental effects, such as systematic errors in interferometer stepping and for which we have no evidence, the most likely explanation is stronger than expected window emission, filling up the troughs between atmospheric lines. We hope to have shortly a better estimation of window emissivity.

Taking now the extreme view that one or other of the experiments was "wrong", what can be said about each? If Q.M.C. have interpreted their data correctly, the turnover of the background is unambiguously demonstrated and we must assume that Berkeley failed to observe the background radiation. On the other hand, if the Q.M.C. dip is an instrumental effect, then the "observed" turnover depends critically on Berkeley's fitting of their model. There are uncertainties both in the line parameters used to construct the model and in the assumptions of the model, particularly those of constant mixing ratio and constant temperature. The properties of the stratosphere are not well known and the Berkeley result could only be accepted provisionally. Confirmation would have to wait upon further measurements of stratospheric emission properties. Q.M.C. will be making laboratory measurements of line strengths shortly, hoping to eliminate at least one uncertainty. It is to be hoped that Leeds' forthcoming flight will provide further evidence which may resolve the discrepancy.

Atmospheric Emission as a Check on the Spectral Measurements of the
Sub-Millimetre Background Radiation – The Leeds Viewpoint
(P. Marchant)

The presence of the residual atmospheric emission spectrum on top of the
sub-mm background spectrum has two aspects. One is that, as pointed out in
our paper (Mercer et al – this volume, p.103) it is very troublesome since
it must be carefully subtracted from the observed spectrum in order to
obtain the background. On the other hand, if one regards the model atmosphere
as correct, and the emission lines, strengths and the column densities of the
dominant emitting species (O_2, O_3 and H_2O) as known, then the atmosphere
serves as an external calibration check. It is this latter aspect which is
discussed here. The validity of the underlying assumptions is discussed in
our manuscript and further justification is shown below by comparison with
the measured spectrum of the Berkeley group (Woody et al[1]).

The Berkeley group, in their balloon experiment, treated the column densities
as free parameters which were used to obtain a best fit to their observed
spectrum. The fitted data yield mixing ratios of 0.234 ± 0.024 and
$5.53 \pm 0.28 \times 10^{-6}$ for O_2 and H_2O respectively (float altitude pressure
= 3.3 mb) and a vertical column density for O_3 of $3.50 \pm 0.18 \times 10^{17}$
molecule cm^{-2}. These are compatible with accepted values (see our paper).

We have produced a computed spectrum (including a 2.99 K Planckian background), in the manner described in our manuscript, using the above atmospheric
parameters and the Berkeley system spectral response. We have then compared
it with the measured spectrum (taken from Figure 1b of Woody et al) obtaining
the agreement shown in Fig. 1 below. Discrepancies are partially due to errors

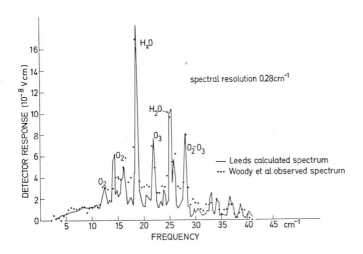

Fig. 1. Comparison of Leeds calculated spectrum with Berkeley
observed spectrum.

in the transcription of the data points (taken at 0.5 cm^{-1} intervals from their graph) and perhaps also due to a different resolution function in the convolution. It should be noted that there is no normalization involved – the intensity scale is absolute – and that the overall agreement is good. This gives us confidence in the validity of the model and the parameters.

A similar calculation was performed with respect to the observations reported in a pre-print of Robson et al[2]. In the pre-print (but not the published report) the composite observed spectrum (resolution = 0.25 cm^{-1}) was displayed. A transcription of this is shown together with our calculated spectrum in Fig. 2. There is a very marked difference between the two. In the pre-print the ordinate scale was given in arbitrary units. For ease of comparison with our calculated spectrum we have normalised both to 1.0×10^{-11} W cm^{-2} sr^{-1} cm at $\bar{\nu} = 6$ cm^{-1} which is approximately the peak of a 2.7 K Planck spectrum. Since window emission and atmospheric emission are both small at 6 cm^{-1}, normalization to the peak intensity is justified. It should be noted that the ordinate scale of the calculated spectrum has been reduced by a factor of 10 with respect to that of the observed spectrum. Also it should be remembered that most ($\sim 80\%$) of a 2.7 K background is below $\bar{\nu} = 10$ cm^{-1}.

The parameters used in the calculation of the Robson et al spectrum were:

$T_{background} = 2.7$ K; $T_{atmosphere} = 250$ K; Pressure = 3 mb; Zenith angle = 50°

$$\text{Mixing ratio } H_2O = 5 \times 10^{-6} \text{ gives } \bar{N}_{H_2O} = 5.0 \times 10^{17} \text{ molecule } cm^{-2}$$

$$\text{Mixing ratio } O_2 = 0.21 \text{ gives } \bar{N}_{O_2} = 2.1 \times 10^{22} \text{ molecule } cm^{-2}$$

$$\bar{N}_{O_3} = 4.7 \times 10^{17} \text{ molecule } cm^{-2}$$

Also included is the emission from the ambient temperature polyethylene window (50 μm thickness). This was taken to have 3.5 times the emission deduced from the bulk sample measurements as given by Chantry et al[3].

The calculation shows a very small ($\sim 1\%$) 2.7 K background contribution to the total spectrum below 45 cm^{-1} which is dominated by atmospheric emission lines (as observed by the Berkeley group). The continuum is due to unresolved atmosphere ($\sim \frac{1}{4}$) and window emission ($\sim 3/4$).

The observed spectrum however exhibits a relatively prominent 'bump' ($\bar{\nu} < 12$ cm^{-1}) which is interpreted by Robson et al as the cosmic background contribution and accounts for about 10% of the total spectrum. The atmospheric line intensities are very much less than we have calculated. Robson et al assume the continuum to be emission from the window and use it for in-flight calibration (although there are indications of anomalous window behaviour below 20 cm^{-1}).

In summary, the observed atmospheric contribution appears to be more than an order of magnitude less (on normalising to 2.7 K Planckian at 6 cm^{-1}) than calculation requires. This is difficult to reconcile with errors in

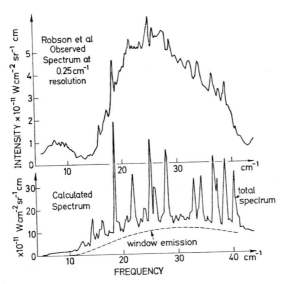

Fig. 2. Comparison of Leeds calculated spectrum with Q.M.C. observed spectrum.

either line strengths or column densities of the emitting molecules, particularly since our calculation for the Berkeley group experiment is consistent with their observed spectrum. Doubt is therefore cast on the validity of the calibration of the Robson et al experiment and consequently on the conclusion that a 2.7 K cosmic background spectrum has been measured at and beyond its peak.

Acknowledgement

We wish to thank Dr. I. Robson for supplying us with the system relative spectral response curve for their experiment so that our calculation could be made.

References

1. Woody, D.P., Mather, J.C., Nishioka, N.S. and Richards, P.L., 1975, Phys.Rev.Lett., 34, 1036.
2. Robson, E.I., Vickers, D.G., Huizinga, J.S., Beckman, J.E. and Clegg, P.E., 1974, Nature, 251, 591.
3. Chantry, G.W., Fleming, J.W., Smith, P.M., Cudby, M. and Willis, H.A., 1971, Chem.Phys.Lett., 10, 473.

Q.M.C. Reply

A general problem encountered in comparing model spectra with experiment is that of matching the resolution of the model to that of the data. Ideally, what should be compared in the case of line spectra is the total power in each line. Unfortunately, particularly in the presence of a strong continuum, the points in an integrated spectrum are not independent, making interpretation difficult. The problem is illustrated in Fig. 1 of the preceeding discussion by Leeds. Whether or not this represents good agreement is a matter of opinion, especially since <u>the parameters used to predict the spectrum are Precisely those obtained from the experimental points (dots)</u>.

Considering now Fig. 2 of the Leeds discussion, the main differences between their predicted and the Q.M.C. observed spectra appear again to be connected with resolution; the Q.M.C. spectrum is clearly not at the computed 1 cm^{-1} resolution. (The scale for the Q.M.C. experiment should be multiplied by a factor of two, since Leeds have fitted the spectrum to 2.7 K at 6 cm^{-1} whereas, in our original publication (Robson et al, 1974) it was pointed out that the measured spectrum was high by a factor of ~ 2 in this region.) In Fig. 7, p.123, we have plotted our measured spectrum (full line) together with our predicted spectrum using the Berkeley column densities and including the Q.M.C. window emission. Both spectra are at 1 cm^{-1} resolution. For this figure, the observed spectrum has been fitted, a little arbitrarily at the <u>peak</u> of the O_3 feature at 21.8 cm^{-1}, representing a compromise between fitting the line peaks and fitting the continuum. Even with this fitting, the spectrum below 10 cm^{-1} is not greatly in excess of the predicted background. (Low frequency oxygen lines below 10 cm^{-1} have not been included in the predicted spectrum.) More realistically, we would fit the total power in the lines which would lower the observed spectrum relative to the model. This will be done when we have obtained a more accurate measure of the window emission.

There remains the deficit of radiation between 10 and 15 cm^{-1}. Much of the predicted power in this region is from ozone, whose concentration is uncertain. The greatest problem is the weakness of the 12 and 14 cm^{-1} oxygen lines; we intend measuring the strengths of these lines in the laboratory.

There is clearly an atmospheric discrepancy between the measurement of Q.M.C. and that of Berkeley, and criticism of the Q.M.C. result has been based on models using the Berkeley data. New data are needed. We look forward with interest to see whether the Leeds experiment will support their predictions.

References

Robson, E.I., Vickers, D.G., Huizinga, J.S., Beckman, J.E. and Clegg, P.E., 1974, <u>Nature</u>, 251, 591.
Woody, D.P., Mather, J.C., Nishioka, N.S. and Richards, P.L., 1975, <u>Phys.Rev. Lett.</u>, 34, 1036.

Other Discussion

<u>Fazio</u> The calculation of atmospheric emissivity at balloon altitudes is extremely important in future design of balloon-borne telescopes: Does anyone have any other results than those presented by Dr. Traub?

<u>Hofmann</u> Dr. Drapak at MPE calculated the emission spectrum of the atmosphere in the infrared region. The result for 35 km altitude is a mean emissivity of 1% at a temperature of 250 K in the spectral range from 20 to 200 μm.

Narlikar Suppose the measurements at $\lambda \lesssim 1$ mm show a turnover as expected
from the blackbody curve, but the background does not decrease rapidly
enough. The question will then arise as to where this extra radiation
originates. If it is anisotropic and of Galactic origin, then other galaxies
are expected to emit similar radiation. Such a radiation from distant
galaxies would be redshifted to $\lambda \gtrsim 1$ mm and would interfere with the
observed blackbody curve. If the extra radiation is isotropic and universal
in origin, the question then is, how does it originate? I am pointing this
out to show that a paradoxical situation will arise if measurements from
$\lambda \lesssim 1$ mm do not show the blackbody form. Observations to date for $\lambda \gtrsim 1$ mm
are not sufficient to establish the blackbody spectrum.

Melchiorri I want to point out that the measurement of the isotropy is
crucial since the 2.7 K cosmic background must be anisotropic, otherwise
the earth is at rest. The observations can be carried out in the radio region,
but ground-based observations are limited by the meteorological systematic
effects which have periods of 12 and 24 hours, exactly the same as the
Doppler effect and isotropic expansion. In my opinion there are two ways of
overcoming this difficulty. The first one, suggested by P. Boynton, consists
in the use of two radiometers operating simultaneously at two different
frequencies in order to use the atmospheric data obtained from one radiometer
to correct the other. We are now studying the best choice of the frequencies
in order to obtain the best correlation.
The second way is to use balloon-borne radio receivers, as in fact is being
done at Princeton.

Chairman's Concluding Remarks

Current knowledge of the long wavelength background radiation ($\lambda > 1$ mm) seems
to agree with a 2.7 K blackbody spectrum. There is considerable inconsistency
and uncertainty in the measurements at shorter wavelengths ($\lambda < 1$ mm),
specifically in the atmospheric models used in the subtraction of the atmospheric emission background. We look forward to the forthcoming results of
the Leeds group; however, any direct and reliable background measurements in
the submillimetre region might become possible only from satellite experiments.

PART 4

LINE ASTRONOMY

THE 4830 MHz H$_2$CO ABSORPTION IN THE DIRECTION OF NGC 6334

F. F. Gardner and J. B. Whiteoak

C.S.I.R.O., Epping, N.S.W. 2121, Australia

Abstract   The distribution of 4830 MHz H$_2$CO absorption has been mapped over the northern radio component, G351.4 + 0.7, of NGC 6334. The absorption with an average velocity of − 4 km s^{-1} has a general association with a dust lane extending across the HII region. In addition there are three H$_2$CO concentrations. Two coincide with OH-emission centres, one of which is an infra-red source. The third, with the highest absorption and smallest angular extent, is centred on a compact continuum radio component; it is also an infrared source.

NGC 6334, a prominent HII region at optical wavelengths, has three main regions of continuum radio emission, identified as G351.0 + 0.7, 351.2 + 0.5, and 351.4 + 0.7 in the 6 cm map made by Goss & Shaver (1970) with a resolution of 4$'$ arc. The northernmost region, G351.4 + 07, is extended to the south-west and with higher resolution breaks up into three components (Schraml & Mezger 1969). Yet another component, near the north-east boundary, is present in the infrared (40 to 350 μm) map of this region (Emerson et al. 1973).

Several molecules have been detected in G351.4 + 0.7. In particular there are two regions of OH emission, NGC 6334 A and B, located at diagonally opposite extremities of the continuum distribution (Gardner et al. 1967). In a previous survey of the 4830 MHz H$_2$CO absorption against continuum sources (Whiteoak & Gardner 1975) six positions in NGC 6334 were investigated. The results suggested an association between the molecular clouds and the lane of dense optical obscuration overlying G351.4 + 0.7. In addition, the highest optical depths were near the OH-emission regions. To investigate the distribution further and to relate it to OH, IR, optical and radio features, a more detailed mapping of the H$_2$CO absorption has been carried out. The details of observation and reduction are described elsewhere (Gardner & Whiteoak 1975). The beam-width was 4 x 4$'$ arc; the position settings are correct to about $\frac{1}{4}'$ arc.

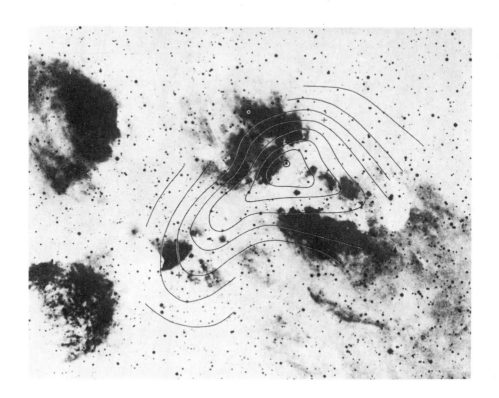

<u>Fig. 1</u>. The 4830 MHz continuum distribution of G351.4 + 0.7
superimposed on a print of NGC 6334. The beam-brightness
temperature of the first contour level and of the contour interval
is 2 K. The crosses show the positions of the OH-emission
regions NGC 6334 A and B.

Figure 1 shows the 4830 MHz continuum distribution of G351.4 + 0.7
superimposed on a print of the Whiteoak extension of the Palomar Sky Atlas.
The continuum maximum is at R.A. $17^h17^m16^s$, Dec. $- 35^o47'$.0 (epoch 1950).
The crosses show the positions of the two OH-emission regions
(R.A. $17^h16^m36^s$.3, Dec. $- 35^o54'57''$, and R.A. $17^h17^m32^s$.1, Dec. $- 35^o44'15''$)
as determined by Raimond & Eliasson (1969).

It is known from the previous investigation (Whiteoak & Gardner 1975) that
the absorption profile of G351.4 + 0.7 contains two main features - a broad
feature centred at a velocity near $- 4$ km s^{-1}, and a fainter, narrow line
with a velocity of 6.5 km s^{-1}. Figure 2 shows the distribution with velocity

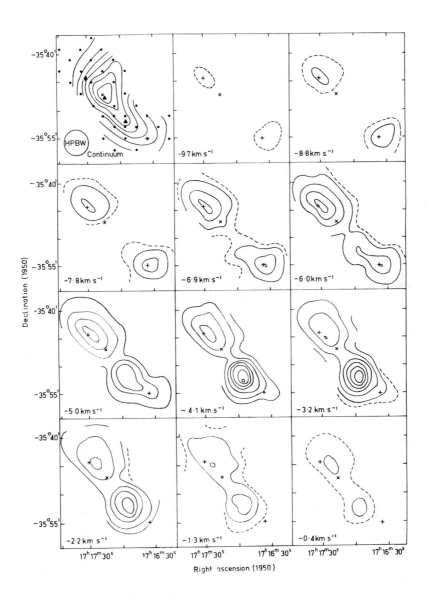

Fig. 2.   The first map shows the continuum distribution of Fig. 1.
Additional features on this map are the maxima in the distribution
of infrared radiation (triangles), the continuum maximum (cross)
and the positions at which observations were made (dots).   The
remaining maps show the distribution with velocity of the H_2CO
absorption associated with the - 4 km s⁻¹ feature.   The beam-
brightness temperature of the first contour level and of the
contour interval (continuous lines) is 0.23 K; for the broken
line contour the level is 0.12 K.

of the absorption associated with the broad feature. Each map is for a
velocity width of ~ 0.9 km s^{-1}. For comparison purposes the first map
shows the continuum distribution. The small dots show the positions at
which observations were made. The crosses show the OH-emission centres
and the continuum maximum. The triangles represent maxima in the distribution
of infrared radiation (Emerson et al. 1973). The right ascensions of the
infrared positions have been increased systematically by 6 s. Such a shift,
which barely exceeds the uncertainty (1' arc) that Emerson et al. assigned
to their positions, seems justified, because there is then a better agreement
in position for the features common to the infrared and radio continuum
distributions.

The 4830 MHz continuum distribution of G351.4 + 0.7 differs markedly from
the distribution of Hα emission, supporting the belief that a lane of dense
obscuration extends south-west to north-east across the source. At infrared
wavelengths the maxima occur near NGC 6334A, near the 4830 MHz continuum
maximum, and at a position along the south-west extension which appears as a
distinct separate component (G351.2 + 0.6) on the map[*] made by Schraml &
Mezger (1969) at a wavelength of 1.95 cm.

The distribution of the absorption in Fig. 2 differs from that of the
continuum, showing that there is not a uniform overlying cloud of H$_2$CO.
It is more elongated in shape, and over narrow ranges of velocity there are
three separate concentrations, none of which coincides with the continuum
maximum. The location and general shape suggest that the absorption is
associated with the dust lane absorbing the Hα emission.

The northernmost concentration is peaked near NGC 6334A (the northern OH
emission centre); it extends in velocity from $- 8.8$ to $- 1.3$ km s^{-1} with
a peak absorption temperature of 2.4 K at $- 4.8$ km s^{-1} (the ratio
absorption-line temperature (T_L)/continuum brightness temperature
$(T_C) = 0.15$). The OH emission velocities, which range between $- 12$ and
$- 9$ km s^{-1} (Robinson et al. 1970) are more negative than those of the

---

[*]This map is displaced from ours by 4 s in right ascension and 1'.6 arc in
declination. Because our positons are considered accurate to $\frac{1}{4}$' arc, we
have modified the 1.95 cm results for the above discussion.

H_2CO absorption.

The southernmost concentration coincides with the southern OH-emission region NGC 6334B. Its absorption extends in velocity from -8.8 to -5.0 km s^{-1} with a maximum value of 1.7 K at -6.5 km s^{-1} ($T_L/T_C = 0.13$). The velocity range is similar to the OH emission (-10 to -4 km s^{-1}).

The most prominent concentration is located between the OH emission regions, about 18 s west and 5' arc south of the continuum peak. Its position coincides with the positions of the southernmost infrared centre and the compact continuum component G351.2 + 0.6 which was mentioned earlier. The absorption attains a peak beam-brightness temperature of 3.8 K ($T_L/T_C = 0.18$). The velocity of peak absorption, -3.6 km s^{-1}, is close to the H109α recombination-line velocity (-3.2 km s^{-1}) of G351.4 + 0.7 (Wilson et al. 1970) and to the velocity (-3.4 km s^{-1}) of the Hα emission of NGC 6334 (Georgelin & Georgelin 1970). The H_2CO absorption has a small angular size - its half-intensity diameter is similar to the beamwidth. This, together with its location near the compact component G351.2 + 0.6, suggests an intrinsic association between component and absorbing cloud.

There is some difficulty in the derivation of optical depths because of the uncertainty in the transition temperature (T_{tr}) defining the relative populations of the 1_{10} and 1_{11} levels involved in the 4830 MHz transition. Because H_2CO emission is only rarely observed, T_{tr} is assumed to be close to the microwave background temperature (T_{bg}) of 2.7 K. However, in the vicinity of NGC 6334A, T_L/T_C exceeds unity, which can be understood only if $T_{tr} < T_{bg}$ (the apparent optical depth τ is related to the ratio $T_L/(T_C + T_{bg} - T_{tr})$). A constant value for T_{tr} of 1.7 K, in keeping with previous results found for cold dust clouds, gives reasonable values of apparent optical depth.

τ is low near the continuum maximum (0.04) but increases substantially to values of 1 and 0.2 near NGC 6334 A and B respectively. Although the greatest line depths occur at the OH-emission regions, the maxima in are displaced 1' to 3' arc further from the position of the continuum peak. This displacement is probably an instrumental effect caused by inadequate angular resolution.

The distribution of the 6.5 km s^{-1} absorption (not shown) is somewhat
similar to the continuum distribution, the result expected if there was an
overlying cloud of fairly uniform density.

The investigation has shown that a molecular cloud containing H$_2$CO with an
average velocity of - 4 km s^{-1} overlies NGC 6334 and is probably
associated with the dust lane crossing the northern Hα emission of the HII
region. The density is low near the maximum of the continuum radio emission,
in agreement with the lower optical obscuration in this direction. The
similarity in the velocities of the cloud, the radio continuum, and the Hα
emission suggests an intrinsic association between cloud and HII region. The
increased absorption near the positions of the OH-emission, infrared components
and a compact continuum component, indicates that these objects are embedded
in dense regions of the molecular cloud where the optical obscuration is very
high. The distributions of H$_2$CO and far-IR (Fig. 2 of Emerson et al. 1973)
are very similar. For NGC 6334A in particular the ratio of IR/radio flux
is very high as pointed out by Emerson et al. If the heating of the dust is
by Lyman continuum then most of the radiation from the star(s) (assuming that
the star has formed) must be absorbed by the dust without contributing to the
ionisation.

References

Emerson, J.P., Jennings, R.E. & Moorwood, A.F.M., 1973, Astrophys.J., 184, 401.
Gardner, F.F., McGee, R.X. & Robinson, B.J., 1967, Aust.J.Phys., 20, 309.
Gardner, F.F. & Whiteoak, J.B., 1975, Mon.Not.R.astr.Soc., 172 (in press).
Georgelin, Y.P. & Georgelin, Y.M., 1970, Astr.Astrophys., 6, 349.
Goss, W.M. & Shaver, P.A., 1970, Aust.J.Phys.Astrophys.Suppl., No.14, 1.
Schraml, J. & Mezger, P.G., 1969, Astrophys.J., 156, 269.
Whiteoak, J.B. & Gardner, F.F., 1974, Astr.Astrophys., 37, 389.
Wilson, T.L., Mezger, P.G., Gardner, F.F. & Milne, D.K., 1970, Astr.Astrophys.,
    6, 364.

DISCUSSION

Emerson Becklin and Neugebauer (Proc. 8th ESLAB Symposium on "HII regions and
the Galactic centre", ESRO SP-105) have detected a very heavily reddened
cluster in the near infrared associated with the northern OH maser which
has high far infrared to radio ratio. They mention some similarities to the
Kleinmann-Low nebula. This makes the region even more interesting and
perhaps suggests that the far infrared emission is predominately from a
molecular cloud.

TIME VARIATIONS AND POSITION JITTER IN INTERSTELLAR WATER MASERS

L. T. Little

Electronics Laboratory, University of Kent at Canterbury, U. K.

Abstract Time variations of intensity in interstellar water masers associated with HII regions, and, in particular, the position jitter observed in components of W49 (H$_2$0) are discussed. It is impossible to account for the position jitter by refraction effects producing "ghost images" of the type suggested by Gold (1968). It is hard to explain time variations of both intensity and position by random motions of the water clouds within the maser, so it seems most likely that changes in excitation are responsible for the observed effects.

1. Introduction

Interstellar water vapour emission has been discovered in more than fifty sources associated with HII regions, often near compact HII components and strong infrared sources. An apparently universal feature of these sources is their variability of intensity, usually on a time scale of a few weeks. As examples Fig. 1 shows spectra of H$_2$0 emission recently discovered near two intense infrared sources AFCRL(UOA) 19 and S140. S140 has recently grown a feature of strength 500 f.u. at + 8 km/s in less than 3 weeks. The – 21 km/s component in AFCRL(UOA) 19 has since disappeared with a time scale < 2 months.

There is no simple relationship between the positions of H$_2$0 sources and those of infrared sources and compact HII regions, although H$_2$0 emission is undoubtedly associated preferentially with them. Figure 2 shows the water sources in W51. The southern source is associated with OH and not coincident with an HII peak. The northern source (Johnston et al. 1973) has no corresponding OH emission, and has usually been omitted from comparative maps of this region. It undoubtedly exists, however, but was presumably too faint to be observed when the interferometer positions of

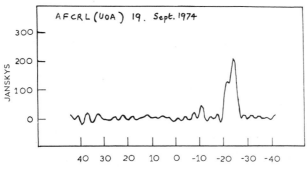

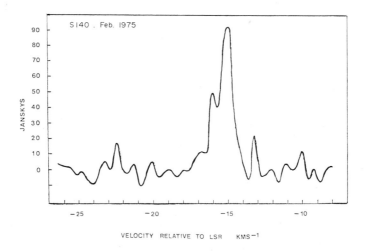

Fig. 1. H_2O emission spectra of AFCRL(UOA) 19 and S140.

Hills et al. (1972) were obtained. Its position has been measured relative to the southern source using a grid technique. This observation, and those of AFCRL(UOA) 19 and S140, are part of a monitoring programme being carried out using the SRC Appleton Laboratory 25 m radio telescope at Chilbolton, Hants., by G. White, E. Parker, G. Macdonald, F. Bale and myself.

2. Position Jitter

Another interesting aspect of water source variability is the position jitter

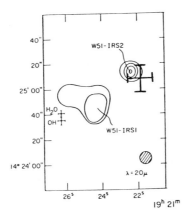

Fig. 2. Position of the W51N water source. The infrared map
is from Wynn-Williams et al., 1974.

which has been observed in components of W49 (H_2O). The relative positions
of components at different radial velocities in this source appear to vary
with time, and it has been suggested (Moran et al. 1973, Baudry et al. 1974)
that a model of an irregular masing cloud put forward by Gold (1968) might
explain the phenomenon. Generally long baseline interferometer observations
show that the strong components appear to be single spots of diameter
$\leq 0''.0005$ arc whose positions jitter around randomly by angles $\sim 0''.05$ arc
in periods of a few months (Knowles et al. 1974). Such behaviour is unlikely
to be due to a number of independent masers varying randomly with time. The
- 7 km component in W49 appeared as a single component at 7 different
positions over a period of 3 years. The probability of seeing one, and
only one, of seven independently varying masers on each of 7 different
occasions is $\sim 10^{-5}$. Neither can it be due to motion of the masing component
itself, since velocities $\sim 10,000$ km/s would be implied (at the distance of
W49) which are much too high.

For these reasons the radiation field expected from an irregular masing cloud
is worth considering; such a cloud might arise in two ways. In the first, all
the irregularities have similar velocities within an overall cloud, in which
case each velocity component would represent a separate cloud. In the second,
there would be a single overall cloud within which there is a considerable
degree of turbulence, producing irregularities in the cloud with a velocity

dispersion similar to that between all the velocity componenets. A particular
component would then arise where the number of irregularities having a
particular radial velocity happened to be large.

3. Can Refraction Effects Produce Position Jitter?

Gold proposed that "ghost images" of an incoherent source at the centre of the
cloud would be formed where the wavefront emerging from the maser was perpendicular to the line of sight. However, this stationary phase condition would
hold only if the source were scanned with a telescope whose beam was much
smaller than the maser. Observations are generally made with interferometers
whose individual elements sum the amplitude and phase over the whole source
and then multiply to give the fringe visibility. Let us suppose that a
distribution of amplitude and phase $\exp\{\phi(\underline{r}) + i \phi'(\underline{r})\}$ exists at the exit
plane of the maser (Fig. 3).

The interferometer receives radiation from a cone defined by the scattering
angle θ_{SC}. The coherence length of this radiation on the ground is
$\xi_C \sim \lambda/\theta_0$ and for the position jitter in W49 typically $\theta_0 \sim 0''.03$ arc
giving $\xi_C \sim 70$ km. Fringes from an interferometer would vary in amplitude
and phase as the earth swept through the coherence pattern at 30 km/s,
yielding a mean fringe visibility which is the Fourier Transform of the
intensity distribution within the cone. The bandwidth of the component (or
receiver) and the finite diameter of the input source also destroy the
instantaneous coherence in a most effective way (if $z^2\theta_0^2/2L > c/\Delta\nu$ for
receiver bandwidth $\Delta\nu$, where L is the thickness of the cloud and z is its
distance from the source, or $z D/L > \lambda/\theta_0$ for source diameter D (see e.g.

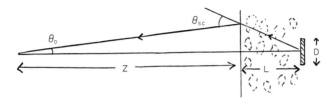

Fig. 3. Simplified maser geometry.

Scheuer 1968). For all reasonable sizes for the maser the radiation must be scattered through a very large angle θ_{SC} and the correlation length of complex amplitude at the exit plane $\sim \lambda/\theta_{SC} \sim L\lambda/z\theta_o$ is extremely small. The correlation length would be 70 cm for an overall source diameter as large as $2''$ arc, which seems very improbable. Within the cone defined by $\theta_o \sim 0''.03$ arc one would expect to see an intensity distribution which varies on this scale, so the overall fringe visibility when observed with an interferometer of fringe spacing $0''.004$ arc would be very low, quite contrary to observation.

Although it seems impossible that refraction effects of the type suggested by Gold could produce the observed position jitter, it is nonetheless interesting to consider the scattering effects which one would expect when radiation contained within a small angle traverses an extended irregular masing medium. When the maser is unsaturated the complex refractive index $n = (1 + n') - i\alpha$ is independent of intensity and a wave travelling a distance dL within the cloud has its complex amplitude changed by $\exp(i\,k\,n\,dL)$. This allows for coherent amplification with a phase change: except in the line centre, where $n' = 0$, $n'\lambda$ and α are of the same order of magnitude (Javan and Kelley 1966). Irregularities in n' and α will scatter the radiation and there is a very strong analogy with the theory of wave propagation in an extended medium containing randomly distributed phase changing irregularities (where $\alpha = 0$). Mercier (1961) has indeed discussed the case where α is negative in connection with an absorbing ionosphere. If the total r.m.s. phase deviation of a ray passing through the maser is ϕ_o' and that of the amplification coefficient is α_o, then the radiation will be scattered through an angle $\theta_{SC} \sim \lambda\,(\phi_o^2 + \phi_o'^2)^{\frac{1}{2}}/\ell$ where ℓ is the typical scale of a masing irregularity. The total gain in nepers required to produce the brightness temperatures observed in water masers is ~ 30, so ϕ_o would be expected to be much less than this. For realistic values of ℓ, θ_{SC} would generally be very small. The correlation length of the intensity variations in the exit plane is $\sim \ell/\phi_o$.

When the radiation is propagating through a medium which it saturates heavily the situation is rather different. Suppose there are N masing irregularities along a ray path. The intensity is proportional to N, so one would expect the variations of intensity across the wave front to be $\sim N^{\frac{1}{2}}$. Hence

$\frac{\Delta I}{I} \sim N^{-\frac{1}{2}}$. Since α and n' are of the same order the phase change $\Delta\phi \sim \Delta A/A \sim N^{-\frac{1}{2}}/2$. For N large the phase change across the emerging wave front will be much less than one radian. This means the angular spectrum of the waves leaving the maser would be dominated by a specular component containing most of the power together with components of relative power $\sim \Delta I/I \sim N^{-\frac{1}{2}}$ scattered into an angle $\sim \lambda/\ell$.

4. Can Random Motion in an Unsaturated Maser Cause the Position Jitter?

If an irregular unsaturated maser amplifies either the microwave background or its own spontaneous emission, then maxima of intensity will be expected along those lines of sight which traverse the largest number of irregularities. Scattering can be neglected if $L \theta_{sc} < \ell$.

One might expect the number of irregularities along a given line of sight to be normally distributed, so the distribution of intensity will be log (normal) and its correlation length ℓ/ϕ_0. The intensity of the maxima will vary on a time scale ℓ/ϕ_0 v where v is a typical random velocity for an irregularity. However, one would expect to see a number of maxima at any one time, rather than a single different one on each occasion (the argument is the same as that employed in Section 2).

5. Can Random Motion in a Saturated Maser Cause the Position Jitter?

Goldreich and Keeley (1972) have shown that the apparent diameter of a saturated spherical maser can be very much less than its true diameter. The radiation appears to come from a small "core" at the centre of the sphere. Within this "core" the total radiation density and hence the saturation is minimised so that rays which pass through it receive large gains compared with rays which do not. On their further progress and amplification out of the sphere rays leaving the "core" receive the greatest part of the available energy at any given point, due to their greater intensity, as a consequence of the equation of radiative transfer. Goldreich and Keeley have suggested that the apparent diameter may be as much as fifty times less than the true diameter. If it is desired to associate observed diameters

$\leq 0''.0005$ arc with the core of a saturated maser whose overall size must
be substantially greater than the position jitter ($\sim 0''.15$ arc), then the
true/apparent diameter ratio must be even larger (> 300). The required
densities of H_2 molecules and the pumping conditions then become rather
unrealistic (e.g. $n_{H_2} = 10^{12}$ cm^{-3} for a 1% inversion).

For a maser in the form of a very thin homogeneous tube, the minimum in the
radiation intensity occurs halfway along its length. Looked at sideways
on this point would give the location of the "core". If the thin tube
were to contain a number N_T of irregularities distributed randomly along
its length, then the point dividing the maser into two parts containing
equal numbers would now define the minimum of the radiation field (Fig. 4a).
If the irregularities are allowed to vary their position and number with
time, this point will vary within $\Delta D \sim D/2N_T^{\frac{1}{2}}$ of the half-way mark, where
D is the overall length. The "core" will be near this minimum and so will
move. A method of increasing the movement would be to remove a central
section of the tube (Fig. 4b). The "core" of the maser would then jump
from side to side depending on which side contained the greater number of
irregularities.

If the irregular distribution of pumped molecules is contained within a
sphere, then its behaviour can only be conjectured at present, since an exact
solution of the radiative transfer equation is a formidable task for such

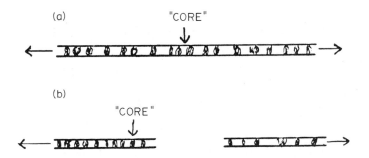

Fig. 4. (a) Thin tube maser containing random irregularities.
(b) With centre section removed.

a system. The distribution of radiation density within the maser is
determined by the distribution of pumped molecules. If this distribution
is changed then the distribution of intensity within the maser will presumably
relax to a new state in a time determined by the light travel time across it
(2.5×10^5 s for a distance 500 A.U. ($\equiv 0''.03$ arc)). It is likely that
only one "core" will be observed for a given configuration since rays passing
through it will saturate the rest of the medium. The position of the "core"
would be near the minimum of the radiation density distribution. The random
configuration of the masing irregularities will change from one independent
state to another in a time scale $T = \ell/v$. For $T = 10^7$ seconds and
$v = 40$ km/s, $\ell \sim 3$ A.U. To explain the position jitter it would have to be
shown that a random change in the configuration of a very large number of
irregularities of this scale could produce a large change in the position of
the intensity minimum (by ~ 500 A.U.).

If we conjecture (by analogy with the thin tube) that this minimum will deviate
from the centre by $D/2N_s^{\frac{1}{2}}$, where D is the diameter and N_s the total number
of irregularities, then for large deviations N_s must be small. The
separation of the irregularities then becomes very much larger than their
sizes and it is hard to see how they might interact to produce a single
"core" of variable position. On the other hand, consider a large sphere
hollow out to a certain diameter. With perfect symmetry the "core" would be
replaced by a "shell" around its inner surface. It the sphere were now
perturbed so that is contained random irregularities then the "shell" might
become unstable and a "core" produced which moves randomly round the inside
surface as the configuration changes.

If the "core" is near the centre of the cloud, random motions of many irregularities along the line of sight will not produce large intensity variations
since $\Delta I/I \sim N^{-\frac{1}{2}}$. The time scale for large changes in intensity is
$T \sim D/v$. For $D > 1000$ A.U. and $v = 40$ km/s, $T > 3 \times 10^9$ s which is much
longer than the observed time scale. These variations would then have to be
attributed to changes in the excitation rather than to the motion of the
water molecules.

6. Discussion and Conclusions

(1) It is impossible to explain the position jitter by refraction effects producing "ghost images" of the type suggested by Gold, or by random motions in an unsaturated irregular maser. A saturated maser is most likely to give the desired appearance of a single spot.

(2) Barring some surprising form of geometry, it seems unlikely that a mechanism of the type suggested in Section 5 could explain the position jitter. Alternatively, some sort of beaming of the pump energy (in the form of ultraviolet or infrared radiation, or conceivably fast particles) might produce changes in the position of greatest excitation. Obscuration by dust clouds moving near a central source might help to produce the beaming.

It is hard to explain time variations of both intensity and position by random motions of the water cloud, so it may prove easier to invoke substantial changes in their excitation. The hot dust-cool gas pumping model of Goldreich and Kwan (1974), in which the driving energy comes from a hot, variable star, depends on such changes.

References

Baudry, A., Forster, J.R., Welch, W.J., 1974, Astron.Astrophys., 36, 217.
Gold, T., 1968, Interstellar Ionised Hydrogen, ed. Y. Terzian, W.A. Benjamin, New York, p.747.
Goldreich, P. and Keeley, D.A., 1972, Astrophys.J., 174, 517.
Goldreich, P. and Kwan, D., 1974, Astrophys.J., 191, 93.
Hills, R., Janssen, M.A., Thorton, D.D., Welch, W.J., 1972, Astrophys.J., 175, L59.
Javan, A. and Kelley, P.L., 1966, IEEE J. of Quantum Electronics, 2, 470.
Johnston, K.J., Sloanaker, R.M., Bologna, J.M., 1973, Astrophys.J., 182, 67.
Knowles, S.H., Johnston, K.J., Moran, J.M., Burke, B.F., Lo, K.Y., Papadopoulous, G.D., 1974, Astron.J., 79, 925.
Mercier, R.P., 1961, Proc.Camb.Phil.Soc., 58, 382.
Moran, J.M., Papadopoulous, G.D., Burke, B.F., Lo, K.Y., Schwartz, P.R., Thacker, D.L., Johnston, K.J., Knowles, S.H., Reisz, A.C., Shapiro, I.I., 1973, Astrophys.J., 185, 535.
Scheuer, P.A.G., 1968, Nature, 218, 920.
Wynn-Williams, C.G., Becklin, E.E. and Neugebauer, G., 1974, Astrophys.J., 187, 473.

DISCUSSION

Gillespie Can the positional variation of these maser sources, found by VLBI, be due to atmospheric fluctuation over one of the telescopes?

Little I don't think so because the positional changes are relative between components. The atmospheric phase differences would be very small when the frequencies are so close together.

Sugden When considering objects which show both H_2O emission to lie closer to the centres of stimulation than the OH emission, as might be expected from the relative energies of the two transitions? In the case of W49(1) a recent absolute position determination at 1667 MHz to $\pm$ 1" arc (done at Jodrell Bank) has shown that H_2O maser emission appears to be coming from inside a larger, clumpy shell of OH emission.

Little I don't think that in general the positions of either OH or H_2O are known with sufficient accuracy to be sure.

Kleinmann Can the positional variations be simply an instrumental effect?

Little The errors in position quoted by Knowles et al are much smaller than the changes.

CARBON MONOXIDE OBSERVATIONS AT 230 GHz

A. R. Gillespie and T. G. Phillips

Queen Mary College, London E.1

Abstract A 230 GHz CO line receiver is described, and preliminary results of observations of M8, M17, W49, W51 and ρ Ophiuchi are presented.

1. Introduction

I intend to describe a 230 GHz (1.3 mm) heterodyne receiver that has been built at Queen Mary College, and show the results of some recent observations of the CO_{2-1} line. The data can be combined with existing CO_{1-0} information to determine the excitation conditions in the regions observed.

2. The Receiver

The system, which is shown in Fig. 1, uses an InSb hot electron bolometer

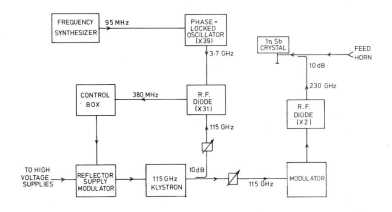

Fig. 1 1.3 mm heterodyne receiver

cooled to liquid helium temperatures (Phillips and Jefferts 1973). When used
as a microwave mixer it samples a part of the source spectrum at the frequency
of the local oscillator. The width of the sample (1.3 km s^{-1}) is set by a
filter. The output from a 115 GHz klystron is frequency doubled using an
R.F. diode to give the local oscillator supply. The klystron frequency is
controlled by a frequency lock loop whose reference comes from the
multiplication of the output of a 95 MHz frequency synthesizer.

The source spectra were obtained by stepping the synthesiser in increments
equivalent to 1 km s^{-1} at 230 GHz. One disadvantage of a frequency
switching system is that the time spent at each velocity is reduced by a
factor corresponding to the number of velocity samples. The system baseline
was determined by making equal numbers of observations at positions away
from the source.

3. The Observations

The observations were made in June of this year, using the 60-in flux
collector at Izana, Tenerife. The beamwidth was 4.5' arc and the system
noise temperature was 2800 K, but this is expected to be improved by a factor
of at least 2 in future observations. The sources observed and the peak
antenna temperatures (T_A^*) corrected for telescope and atmospheric losses are
given in Table 1.

TABLE   1

	Sources Observed	T_A^* (K)
HII Regions	M8	23
	M17	36
	W49	14
	W51	14
Dark Cloud	ρ Ophiuchi	21

The spectra obtained for the four HII regions are shown in Fig. 2. The
profiles and brightness temperatures are in good agreement with the CO$_{1-0}$

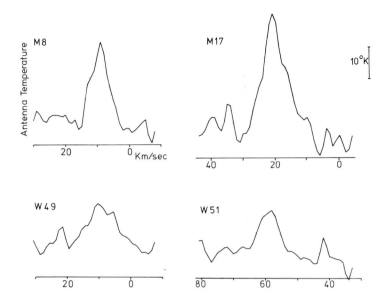

Fig. 2 CO_{2-1} scans of HII regions

data for these sources, confirming that the $^{12}C^{16}O$ is optically opaque and
in thermal equilibrium with the molecular hydrogen in these areas.

The dark cloud in the ρ Ophiuchi complex was observed at the position of the
CO_{1-0} peak detected by Encrenaz (1974).   The CO_{2-1} profile is shown in
Fig.3, superimposed on Encrenaz's CO_{1-0} profile, and again there is good
general agreement, except for the feature at 5 km s^{-1} in the CO_{1-0} data
which is not obvious at 230 GHz.   There is, however, an asymmetry in the
CO_{2-1} spectrum.   It may be concluded that the ρ Ophiuchi cloud is similar
in properties to the molecular clouds surrounding the HII regions.

References

Phillips, T.G., and Jefferts, K.B., 1973, Rev.Sci.Instrum, 44, 1009.
Encrenaz, P.J., 1974, Astrophys.J., 189, L135.

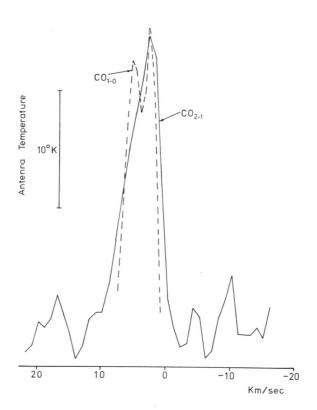

Fig. 3 CO_{2-1} and CO_{1-0} scans of Rho Ophiuchi.

DISCUSSION

Mezger   Is the spatial structure found in the 1.3 mm line in the Orion cloud
still valid and has similar structure been found in other clouds?
Gillespie   Yes.
Mezger   The similarity of the excitation temperature in ρ Ophiuchi to that
in clouds associated with HII regions is not too surprising. Observations
of the C-recombination lines have shown the existence at least of B-stars,
which are responsible for the heating of the dust.
Gillespie   I agree that the Carbon recombination lines have shown the
presence of exciting stars, but these observations are the first direct
evidence that the molecular clouds are in thermal equilibrium in a dark
cloud.
Mezger   Does the similarity of the 2.6 and 1.3 mm CO lines mean that opacity
effects in the J = 1-0 line have been overestimated?
Gillespie   Yes.
Gardner   Do you consider that the +35 km s^{-1} emission in M17 is real?
Gillespie   We have not yet analyzed the data sufficiently to give a firm
answer, as it is only a few times the noise level.

HIGH SPECTRAL RESOLUTION LINE OBSERVATIONS IN THE INFRARED

M. Anderegg, A. F. M. Moorwood and H. H. Hippelein[*]

Astronomy Division, ESTEC, Noordwijk, Holland

J. P. Baluteau and E. Bussoletti[†]

Groupe Infra-rouge Spatial, Observatoire de Meudon, France

and

N. Coron

Laboratoire de Physique Stellaire de Planetaire, Verriere-le-Buisson, France

Abstract A high resolution ($\lambda/\Delta\lambda \simeq 10^4$) Michelson interferometer
has been built for use on NASA's AIRO (C141) telescope to study
far infrared emission lines from HII regions. The instrument
operates in the rapid scanning mode under computer control and
spectra are computed, averaged and displayed on line. In
preparation for the first airborne measurements the instrument has
recently been used on the 1 m telescope at Observatoire du Pic du
Midi to search for SIV (10.5 µm) and SIII (18.7 µm) line emission
from the Orion nebula and from a number of planetary nebulae.
Atmospheric emission spectra were also obtained between 17.5 µm
and 20 µm at a spectral resolution of 0.02 cm^{-1}. Neither of the
sulphur lines appears to have been detected and for the Orion
nebula the upper limits obtained are lower than the latest
theoretical predictions. Analysis of all the data is not yet
complete however and this is a preliminary result based on only a
sample of the spectra obtained.

1. Introduction

We describe briefly a high resolution rapid scanning, Michelson
interferometer which has been built for airborne measurements of far
infrared line emission from HII regions. The first flights with the
instrument on NASA's AIRO (C141) flying observatory are planned for November
1975 but spectra have recently been obtained from the ground using the 1 m

[*] *On leave from Max-Planck-Institut fur Astronomie, Heidelberg, W. Germany*

[†] *Now at Istituto di Fisica, Universita di Lecce, Italy.*

telescope at Observatoire du Pic du Midi.   Observations were made in both the 10 μm and 18 μm atmospheric windows to search for SIV (10.5 μm) and SIII (18.7 μm) line emission from the Orion nebula. The planetary nebulae IC418, NGC6572 and NGC3242 plus the moon were also observed. Reduction of the data is not yet complete but preliminary analysis of the Orion spectra indicates that both sulphur lines are weaker than theoretically predicted. Most of the astronomical observations were made with a spectral resolution of around 0.1 cm^{-1}. Atmospheric emission spectra between 17.5 μm and 20 μm were recorded however at the maximum instrumental resolution of 0.02 cm^{-1} and are, to our knowledge, the highest resolution observations available in this region.

2. The Instrument

2.1. Concept

Two major considerations led to the choice of a Michelson interferometer for the airborne measurements. Firstly the lines of interest are distributed throughout the infrared from a few microns to several hundred microns. Secondly, many of the lines are close to atmospheric lines which are strong even at aircraft altitudes. The Michelson interferometer offers the advantages that it can be used throughout the spectral range of interest and the resolution which can be achieved is sufficient to resolve the atmospheric lines. In addition, the multiplex and throughput advantages are also important in view of the low intensities of the lines and the fact that most HII regions are extended on a scale of several arc minutes. The rapid scanning mode was chosen to avoid effects due to signal variations caused by scintillation or pointing fluctuations and to minimize the periods during which continuous stable telescope pointing is required on the aircraft. Operation in this mode also allows us to dispense with sky chopping which creates difficulties when observing extended sources. The scan rates selected produce fringe frequencies at around 80 Hz and the high spectral resolution is used to discriminate against atmospheric lines and to maximize visibility of the astrophysical lines against the continuum.

2.2 Optical Design

The optical arrangement for operation on the C141 is shown schematically in

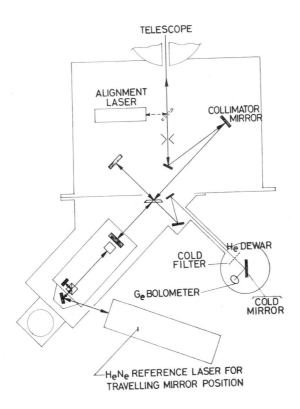

Fig. 1. Optical configuration of the interferometer.

Fig. 1. Input collimation is by an off axis spherical mirror which sets the
limiting resolution (defined as $R = 2\pi/\Omega$) at 2×10^4 for a 1' arc aperture stop
and the 1 m AIRO telescope. The beamsplitter is a stretched Mylar film, the
thickness of which is selected for the particular wavelength region of interest.
Mylar has been used so far at wavelengths as short as 10 μm but it is not
ideal below 20 μm and it is intended in future to use a solid, coated ZnSe
beamsplitter at the shorter wavelengths. The moving mirror is carried on a
high precision translation table which is capable of introducing a maximum
path difference of 30 cm, equivalent to a resolution of better than 0.02 cm^{-1}.
Mirror position is measured to 0.1 μm by the HeNe laser interferometer which is
also used for servo-control of the scanning speed to within 0.1% up to a
maximum speed of 2 cm^{-1} (path difference). The actual mirror translation speed
is selected, depending on wavelength region, to give fringe frequencies in the

range 60–80 Hz.　Higher frequencies become impracticable due to limitations set by the bolometer response and the sampling time required to achieve 16 bit digitization accuracy.　Laboratory tests using a 10.6 μm CO_2 laser as source have verified that the apparatus function is symmetrical and that no 'ghost' lines due to periodic sampling errors are present in the spectrum. The detector is a Coron 'three parts' bolometer which is mounted on the cold plate of a liquid He cryostat together with the final focussing mirror and optical filters.　N.E.P. is less than 5×10^{-14} W $Hz^{-\frac{1}{2}}$ at the operating temperature around 1.5 K　and to reduce the thermal background noise below this figure it is necessary to limit the width of the cold filter for any measurement to around $\Delta\lambda \simeq 0.1 \lambda$.

Alignment of the interferometer and alignment of the interferometer with the telescope is carried out using the second HeNe laser.　A double-prism beamsplitter can be inserted remotely into the optical path and the laser beam is divided such that one beam travels to the centre of the telescope secondary mirror and back while the other passes through the interferometer. For source acquisition a two position mirror (not shown in Fig. 1) is moved to allow viewing of the focal plane through the guiding eyepiece.　In its normal position the beam passes through a hole in this mirror but the surrounding field can still be viewed to check for any guidance drifts during the measurements.　A variable temperature blackbody can also be moved into the beam for calibration and a rotating chopper situated in front of this is used for alignment of the detector with the entrance aperture at the telescope focus.

2.3　Electronic Control and Data Handling.

The interferometer is operated on-line with an HP2100-A computer which both controls and monitors instrument performance and performs on-line Fourier transformation and spectrum display.　All data are also recorded on magnetic tape for later off-line processing on the large computers at ESTEC and Meudon.

Parameters such as scan length (resolution), mirror speed and sampling interval (wavelength region) are selected manually through switches on the control unit.　Operation is then controlled via the computer which receives mirror position and direction information enabling it to transmit the commands for terminating the scan and initiating rapid return of the mirror to its start position. Fourier transformation of the interferograms is carried out during the few seconds taken for rapid return of the mirror. The sequence is then either terminated or repeated depending on the mode selected. Error messages are generated in the case of instrument malfunction or if the number of data words received is incorrect.

Sampling of the interferograms at equal path difference intervals is triggered by pulses derived from the laser interferometer. Digitization is to 16 bits after the detector output has first been amplified by a low noise (< 4nV/Hz at 80 Hz) preamplifier followed by a variable gain AC amplifier and 50-120 Hz bandpass filter. The interferograms are single sided except for a short section recorded beyond the zero phase point. This section is used during on-line operation to compute any phase error due to drift in the zero phase point and after each scan the sampling positions are automatically adjusted to follow any such change. During off-line processing the short double sided section is used for full phase correction of the interferograms. The total number of data points/scan can be up to about 50,000 but as the spectral range is limited by the cold optical filter, the interferograms can be digitally filtered and downsampled to a more manageable number of points prior to on-line Fourier transformation. The digital filter used at present involves 128 point averages and the downsampling ratio is ten. At the end of each scan the spectrum and then the running average spectrum are displayed on a Tektronix visual display terminal and hard copies can be made as desired. Following inspection of each spectrum displayed the operator can delete it from the average if necessary. The on-line spectra are invaluable for assessing instrument performance and the integration times required. To achieve the full resolution and dynamic range capabilities however and to perform phase correction and apodization it is necessary to recompute the spectra later on a large digital computer.

3. Observations at Pic du Midi

Although intended for airborne observations, valuable experience has been gained recently with the instrument on the 1 m telescope at Observatoire du Pic du Midi. The observatory is at an altitude of nearly 3000 m in the Hautes Pyrenees, France. The observations described here were made during

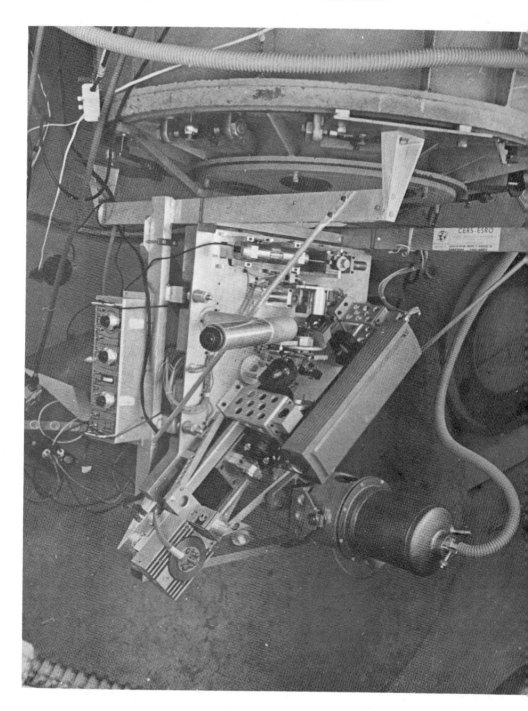

Fig. 2. The instrument mounted at the Cassegrain focus of the
         1 m Pic du Midi telescope.

our second visit, at the end of February 1975. Weather conditions were less than ideal but a few nights were clear enough and the atmospheric transmission high enough to make sensible observations. During our first visit in January only 18 μm measurements were made and atmospheric transmission was extremely low. Fig. 2 is a photograph of the instrument mounted at the Cassegrain focus of the telescope. Apart from improvising the mounting bracket the only major changes necessary from the planned airborne configuration were the repositioning of the HeNe laser over the interferometer to reduce the overall length and the addition of the guiding eyepiece.

The principal observations made were those of the Orion nebula to search for the so far undetected lines of SIV at 10.5 μm and SIII at 18.68 μm. Three planetary nebulae were also observed (IC418 at 18.68 μm and NGC 6572 and NGC 3242 at 10.5 μm) but the telescope is rather small for planetary nebulae measurements unless extremely long integration times can be achieved. The moon was also observed at 10.5 μm as an independent calibration check. In each case, background spectra were recorded with comparable total integration times to the source spectra. Atmospheric emission spectra were also recorded between 17.5 μm and 20 μm at the maximum instrumental resolution of 0.02 cm^{-1}. The important observational parameters are summarized in Table 1.

TABLE 1.   OBSERVATIONAL PARAMETERS - PIC DU MIDI

Telescope Diameter	1m	
Beam Width	50"	
Centre Wavelength	10.6 μm	18.7 μm
Spectral Bandwidth	$\simeq$ 0.5 μm	$\simeq$ 2 μm
Resolution	0.16 cm^{-1}	0.1 cm^{-1} (0.02cm^{-1})
Integration Time/Scan	40 s	30 s (150 s)

Analysis of the data obtained is not yet complete. This is mainly because the software necessary is being developed concurrently with the analysis and the usual frustrations are being encountered. In particular none of the spectra so far computed have been calibrated to remove the instrument response. An approximate intensity calibration has been made however to enable a preliminary assessment of the noise performance to be made. The r.m.s. noise on the high resolution (0.02 cm^{-1}) atmospheric spectra is equivalent to a signal of approximately 2.5×10^{-17} W cm^{-2} per spectral element in the centre of the band compared with the theoretically predicted value of 0.5×10^{-17} W cm^{-2}. This factor of five disagreement is not considered too serious in view of the uncertainties in the quantities (e.g. overall optical transmission) entering into the theoretical calculation.   It is also possible that these measurements were photon noise limited. Steps are being taken now however to optimize the system performance before the airborne observations. The particular spectrum selected for the above comparison was recorded after midnight. The significance of this is that most of the Orion spectra were recorded before midnight when an extra source of noise was present in the form of R.F. interference from a high powered T.V. transmitter which is situated only a few hundred metres from the observatory. The increase in noise was typically between two and five depending on the observing direction relative to the T.V. mast. Another variable noise contribution resulted from vibration of the beamsplitter due to the extremely high wind which accompanied almost all our clear observing periods. This effect was particularly noticeable at 10 μm. The interferometer will be evacuated on the C141 but this was not possible at Pic du Midi. One consequence of the two extra noise sources is that the optimum signal to noise ratio is not achieved by a straightforward averaging of all the interferograms recorded on a particular source. A procedure is now being developed therefore whereby 'noisy' interferograms can be easily identified and rejected from the average.   With hindsight this could have been carried out more efficiently during on-line operation if only part of the spectrum had been displayed to increase the dynamic range.

In the following section we discuss the preliminary results already obtained from an analysis of the Orion observations and present a high resolution spectrum of the atmosphere between 17.5 μm and 20 μm.

4. Preliminary Results

4.1 Atmospheric Emission in the 18 μm Window.

Figure 3 is an atmospheric emission spectrum recorded at an elevation of 63°

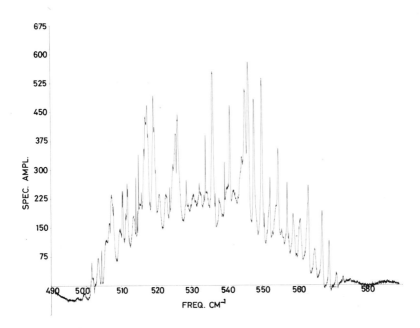

<u>Fig. 3.</u>   Atmospheric emission in the 18 μm 'window' recorded at an
elevation of 63°.   Spectral resolution is 0.02 cm^{-1} and
the recording time was 150s. No correction has been made
for the instrumental response.

and at the maximum instrumental resolution of 0.02 cm^{-1}. This spectrum was
computed on the ESTEC ICL 470 computer from a single interferogram recorded
in 150 s. The apparent baseline error is due to the fact that the
interferogram was not phase corrected before Fourier transformation.

Several of the water vapour lines are saturated and some show self absorption
in the instrumental air path. Line positions have been checked against the
AFCRL Atmospheric Absorption Line Parameters Compilation (McClatchey et al.
1973) which quotes positions within 0.02 - 0.05 cm^{-1} in this region. Based
on a comparison of twelve unsaturated water vapour lines distributed
throughout the spectrum, the measured positions are shifted to lower wave-
numbers by an amount of 0.07 ± 0.03 cm^{-1} with respect to the AFCRL compilation.
The most likely explanation for such a shift is that the reference laser

interferometer was slightly misaligned with respect to the direction of the scanning mirror. As the instrumental response has not yet been removed, no detailed analysis of the spectrum has been carried out. A provisional estimate however gives a figure of around 1 mm for the precipitable water vapour in the path.

4.2 Observations of the Orion Nebula around 18.7 μm.

The SIII doublet lines at 18.68 μm (535.3 cm^{-1}) and 33.65 μm (297.2 cm^{-1}) should be among the brightest infrared lines in the Orion spectrum and are of interest because their intensity ratio is a sensitive measure of electron density beyond Ne $\simeq 10^3$ cm^{-3}. No measurements have been reported of either line. The line predicted to be at 535.3 cm^{-1} lies just to the side of a saturated water vapour line centred at 536.2 cm^{-1} but should be observable from the ground at a dry site. Atmospheric transmission at this position was around 70% during our second observation period. Six regions in the nebula were observed. No line appears in the spectra displayed during the measurements but to date only the spectra of the trapezium region have been analysed down to the noise limit. Figure 4 is an average of 39 spectra at a resolution of 0.1 cm^{-1} in the trapezium direction. Figure 5a is an expanded plot of the spectrum around the predicted SIII line position and the error bar shown represents one standard deviation of the noise recorded just outside the optical bandpass. Based on our preliminary blackbody calibration this noise is equivalent to a signal of 3×10^{-17} W cm^{-2}/spectral element in a 50" beam. Figure 5b is the difference between the source spectrum and the background spectrum and one standard deviation of the noise corresponds to a signal in one spectral element of about 7×10^{-17} W cm^{-2}. The background spectrum was only averaged over 20 scans and the noise is higher than on the source spectrum. An average over 40 scans is now being computed. No signal in excess of three standard deviations or 21×10^{-17} W cm^{-2} appears in the spectrum. This is equivalent to a line intensity from the whole nebula of $\simeq 4\times10^{-15}$ W cm^{-2} which is less than both the value of 2.1×10^{-14} W cm^{-2} predicted by Petrosian (1970) and the more recent value of 1.2×10^{-14} W cm^{-2} due to Simpson (1975).

Uncertainties in our own preliminary calibration and in the parameters such as abundance and collision strength which enter into the theoretical

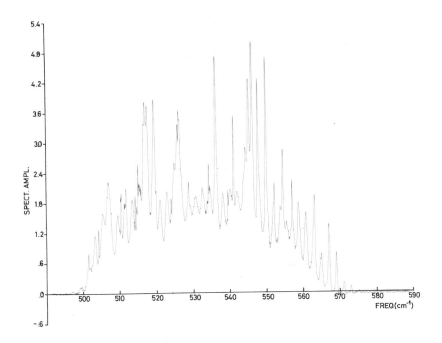

Fig. 4.   Average of 39 spectra recorded while pointing to the
          trapezium region of the Orion nebula.   Spectral
          resolution is 0.1 cm^{-1}.

calculations are sufficient to explain such a discrepancy.   It is also
possible however that the line position is in error and that the emission
suffers significant absorption in the atmosphere.   We shall attempt to
reduce the limiting noise in the spectrum already obtained by adding the
remaining scans not yet included and by rejecting the noisiest records.
This line will also be included in the programme of airborne observations on
the C141.

4.3   Observations of the Orion Nebula around 10.5 μm.

The line of SIV close to 10.5 μm is predicted to be the strongest emission
line from the Orion nebula but no measurements of it have been reported.   The
line has been detected in several planetary nebulae and the highest spectral
resolution observations available are those of Holtz et al. (1971) who

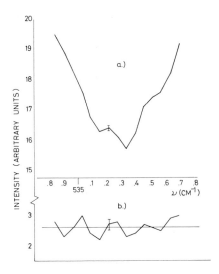

Fig. 5.   a) Portion of the spectrum (Fig.4) around the predicted position of the SIII line (535.3 cm^{-1}). The error bar is one standard deviation of the noise and corresponds to a signal in one spectral element of about 3×10^{-17} W cm^{-2}/beam.

b) Difference of above spectrum and background spectrum. The error bar is one standard deviation and corresponds to a signal in one spectral element of about 7×10^{-17} W cm^{-2} /beam.

measured the rest frequency as 951.5 cm^{-1}. The previous best estimate of the line position was 950.3 cm^{-1}, determined from differences in ultraviolet transitions.

Figure 6 is the source-background difference spectrum obtained on the trapezium region during one set of observations. The source and background spectra were each computed as the average of 24 scans and the spectral resolution is 0.16 cm^{-1}. There are no statistically significant features in the region of interest. The largest peak is at 949.63 cm^{-1} and corresponds to an intensity of $(24 \pm 8) \times 10^{-17}$ W cm^{-2} / spectral element in a 50" beam or $\simeq 5.5 \times 10^{-15}$ W cm^{-2} from the whole nebula. This is again lower than the predicted intensities due to Petrosian (10^{-13} W cm^{-2}) and Simpson (1.4×10^{-14} W cm^{-2}).

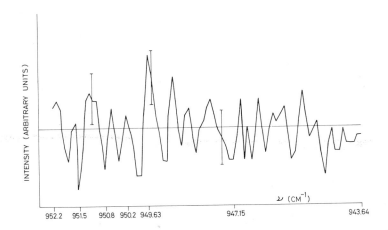

Fig. 6. Orion trapezium source-background spectrum around the position of the SIV line. Resolution is 0.16 cm^{-1} and the one standard deviation error bar corresponds to a signal in one spectral element of about 8×10^{-17} W cm^{-2}/ beam.

5. Summary and Conclusions

Spectra around 10.5 μm and 18.7 μm have been obtained from a high altitude observatory with a high resolution Michelson interferometer intended for airborne measurements of far infrared emission lines. The maximum spectral resolution of 0.02 cm^{-1} was used to measure the emission spectrum of the atmosphere between 17.5 μm and 20 μm. Observations at resolutions around 0.1 cm^{-1} were also made of the Orion nebula but preliminary analysis of the spectra reveals no lines of the predicted intensity at the positions of the SIV and SIII transitions. A final result for those observations is not yet available. As these were the first measurements made with the instrument, the spectral resolution and noise performance achieved are considered to be highly satisfactory. It is expected however that the sensitivity will be improved before the first airborne observations on NASA's C141 aircraft.

Acknowledgments

As people working at three separate institutes have and are contributing in various aspects of this programme, it is impossible to acknowledge individually all those people who have helped both directly and by their support in the development of the instrument. Special thanks must go to M. J. Gilles, however, who has worked tirelessly to develop and run the computer programmes, J.E. Beckman who has contributed actively at various times and all the hardworking technical staff without whose specialist skills this project would still be an idea. Our thanks also to the Director and staff of the Pic du Midi Observatory for their help and support during our two visits there.

References

1. Holtz, J.Z., Geballe, T.R., and Rank, D.M., 1971, Astrophys. J. 164, L29.
2. McClatchey, R.A., Benedict, W.S., Clough, S.A., Burch, D.E., Calfee, R.F., Fox, K., Rothman, L.S., and Garing, J.S., 1973, AFCRL Atmospheric Absorption Line Parameters Compilation, AFCRL-TR-73-0096.
3. Petrosian, V., 1970, Astrophys.J., 159, 833.
4. Simpson, J.P., 1975, Astr. Astrophys., 39, 43.

DISCUSSION

Sollner Would you care to comment on the Pic du Midi as an observatory for infrared observations which are sensitive to atmospheric water vapour?
Moorwood We only have the limited experience of two visits, one in January and one in February 1975. Conditions were generally bad on both occasions but during the second visit, atmospheric transmission around 18 μm was high on several nights. Comments I have had from other people indicate that water vapour above the Pic is very variable but that good nights can be very good.
Robson We found (using the 105-cm telescope) the precipitable water vapour content quite variable from the Pic du Midi, but it is interesting to note the lack of correlation between precipitable water vapour and ground level humidity. We have found the same effects from Mr.Evans and Mauna Kea observations.

ON THE SENSITIVITY OF HETERODYNE DETECTORS IN FAR INFRARED ASTRONOMY

H. G. van Bueren

The Astronomical Institute, Utrecht, The Netherlands

Abstract   The signal-to-noise ratio of astronomical heterodyne detection infrared spectrographs is considered, taking into account background, linewidth and seeing effects. A comparison with incoherent detector systems is presented.

1. Incoherent detector

The signal-to-noise ratio of an incoherent, photo-emitting diode detector system placed behind a telescope, is given by (see e.g. De Graauw, 1975)

$$SNR_{(inc)} = \frac{I_s^2}{2eB_{(inc)} (I_s + I_b + I_D + I_A)} . \tag{1}$$

Here I_s and I_b are the photocurrents generated by the signal and by the background, respectively. I_D is the detector dark current and I_A the noise-equivalent current of the amplifier. $B_{(inc)}$ is the noise-equivalent bandwidth of the latter and e stands for the electronic charge. (For a photoconductive detector recombination noise introduces an extra factor 2 in the denominator). Between the incident radiation powers P_s and P_b (in watts) and the corresponding currents I_s and I_b (in amps) the relations hold

$$I_s = \frac{\eta e}{h\nu} P_s , \tag{2}$$

and

$$I_b = \frac{\eta e}{h\nu} P_b ,$$

where η is the quantum efficiency of the detector material at the average radiation frequency ν; h is Planck's constant. It is always possible in a

restricted frequency range to express a radiation power in a temperature
with the aid of Planck's law. If the telescope aperture is A and it
accepts a solid angle Ω, and if the solid angle subtended by the source is
$\Omega_s \leq \Omega$, then

$$P_s = 2 \frac{h\nu^3}{c^2} A \Omega_s \Delta\nu_s \delta_s \tag{3}$$

and

$$P_b = 2 \frac{h\nu^3}{c^2} A \Omega \Delta\nu_b \delta_b$$

Here $\Delta\nu_s$ and $\Delta\nu_b$ are the received radiation bandwidths of source and
background, respectively. The 'Planck factors' δ are given by

$$\delta_{s,b} = \frac{1}{e^{h\nu/\kappa T_{s,b}} - 1} . \tag{4}$$

Assuming that the background-induced noise dominates over all other noise
sources, after inserting equations (2) and (3) into Eq.(1), we find

$$SNR_{(inc)} = \eta \frac{A}{\lambda^2} \frac{\Omega_s^2 \Delta\nu_s^2 \delta_s^2}{\Omega B_{(inc)} \Delta\nu_b \delta_b} \tag{5}$$

where we have introduced the average wavelength $\lambda = \frac{c}{\nu}$.   By putting the
left hand side of Eq.(5) equal to unity we obtain the noise-equivalent power
(NEP) of the system. Since we are primarily interested in the source
radiation characteristics, which are expressed by the factor δ_s, we just
write

$$\delta_{s(inc)}^{min} = \{\frac{B_{(inc)} \Delta\nu_b}{\Delta\nu_s^2} \frac{A_c}{A} \frac{\Omega \delta_b}{\Omega_s \eta}\}^{1/2} \tag{6}$$

where $A_c = \frac{\lambda^2}{\Omega_s}$ is the coherence area of the source. This equation is
completely equivalent to the more commonly used expression

$$P_{s(inc)}^{min} \; (= NEP) = \{\frac{2h\nu \, P_b \, B_{(inc)}}{\eta}\}^{1/2} \tag{7}$$

H.G.VAN BUEREN

2. Coherent detector

The signal-to-noise power ratio of a coherent (photo-diode) detector is given by (De Graauw, 1975):

$$SNR_{(coh)} = \frac{I_s{}^* I_L}{eB_{(coh)}(I_s + I_b + I_D + I_A + I_L) + (I_s{}^*I_b{}^* + I_b{}^*I_L)} , \qquad (8)$$

where I_L stands for the laser-induced photocurrent and asterisks denote that the signal and background radiation powers are integrated over twice the effective intermediate frequency bandwidth $B_{(coh)}$ on both sides of the laser frequency ν_L, to yield the corresponding currents. Assuming the laser-induced current to dominate over all other currents, one has

$$SNR_{(coh)} = \frac{I_s{}^*}{eB_{(coh)} + I_b{}^*} \qquad (9)$$

(Again, for photoconductive detectors a factor 2 should be introduced before the factor $eB_{(coh)}$).

One has according to Eq.(3), and taking polarization into account:

$$P_s{}^* = \frac{h\nu^3}{c^2} A \Omega_s \Delta\nu_s{}' \delta_s , \qquad (10)$$

where the optical bandwidth $\Delta\nu_s{}'$ is now limited to $2B_{coh}$; if it is larger, $\Delta\nu_s{}'$ has to be replaced by $2B_{(coh)}$ because that is the maximum detectable optical bandwidth. Further we have taken $\Omega_s \le \lambda^2/A$; again, if it is larger, Ω_s should be replaced by λ^2/A, because the etendue $A\Omega$ of a coherent detection system is equal to λ^2. In view of this latter limitation we can write for the integrated background power

$$P_b{}^* = \frac{h\nu^3}{c^2} \lambda^2 2B_{(coh)}\delta_b = 2 h\nu B_{(coh)} \delta_b. \qquad (11)$$

Using Eq.(2) and the definition of the coherence area A_c given earlier, we finally obtain from Eq.(9):

$$\text{SNR}_{(coh)} = \frac{A \, \Delta\nu_s{}'}{A_c \, B_{(coh)}} \, \frac{\eta \, \delta_s}{1 + 2\eta \, \delta_b} \, . \tag{12}$$

Practically the same formula was derived earlier by Cummins and Swinney(1970). Again, by putting the left hand member equal to unity, we obtain the noise-equivalent power. Written in a way comparable to Eq.(6) we have

$$\min_{\delta_{s(coh)}} = \frac{A_c}{A} \, \frac{B_{(coh)}}{\Delta\nu_s{}'} \, (2\delta_b + \frac{1}{\eta}) \, . \tag{13}$$

3. Comparison

We can now compare the expression (6) with (13).

Dividing Eq.(13) by Eq.(6), we obtain a dimensionless ratio r indicating the relative merits of coherent and incoherent detection. For coherent detection to be favourable, this ratio should be less than unity, and in general as small as possible. According to the considerations of the previous section one has to distinguish between several cases:

a) small (point like) source $(A_c \geq A)$ with narrow spectral line $(\Delta\nu_s{}' \leq B_{coh})$:

$$r = (\frac{\lambda^2}{A\Omega})^{\frac{1}{2}} \, \frac{B_{(coh)}{}^2}{(B_{(inc)}\Delta\nu_b)^{\frac{1}{2}}}^{\frac{1}{2}} \, . \, \frac{\Delta\nu_s}{\Delta\nu_s{}'} \, \{(2\eta \, \delta_b)^{\frac{1}{2}} + (2\eta \, \delta_b)^{-\frac{1}{2}}\} \tag{14a}$$

b) small source $(A_c \geq A)$ with wide spectral line or continuous spectrum $(\Delta\nu_s{}' \gg B_{coh})$:

$$r = (\frac{\lambda^2}{A\Omega})^{\frac{1}{2}} \, . \, \frac{\Delta\nu_s}{(2B_{(inc)}\Delta\nu_b)^{\frac{1}{2}}} \, \{(2\eta \, \delta_b)^{\frac{1}{2}} + (2\eta \, \delta_b)^{-\frac{1}{2}}\} \, , \tag{14b}$$

c) and d)

narrow and wide spectral line, respectively, but extended source $(A_c < A)$. In this case both above expressions have still to be multiplied with the factor $\frac{A}{A_c}$, unless the telescope is diffraction limited.

Assuming $B_{(coh)}$ and $B_{(inc)}$ to be of the same order of magnitude B, a
point source, and the complete instrument to be diffraction limited
$(A\Omega = \lambda^2)$, we can simplify the expressions (14) to

$$r = \frac{\Delta\nu_s}{(2B\ \Delta\nu_b)^{\frac{1}{2}}}\ [\frac{2B}{\Delta\nu_s '}]\ \{\ (2\eta\delta_b)^{\frac{1}{2}}\ +\ (2\eta\delta_b)^{-\frac{1}{2}}\ \} \tag{15}$$

where the factor between square brackets has to be taken into account only
for very narrow lines $(\Delta\nu_s ' < B_{(coh)})$.

To take the effect of seeing into consideration, we have to replace in
first instance the full telescope area A in Eq.(13) by the area of the
seeing disc A_t, if the latter is smaller. The value of A_t is in general
a function of wavelength, it is probably of the order of a few m^2 in the
far infrared. A similar replacement is not necessary in Eq.(6), since
direct detection is insensitive to phase distortions of the wavefront.
Therefore, in Eq.(14) and Eq.(15) a proportionality factor $[\ A/A_t]$ (if $A > A_t$)
has still to be added to account for seeing effects.

4. Discussion

From Eq.(15) it follows that under ideal comparison conditions, coherent
detection has its greatest advantages over direct detection if the background
radiation power lies within certain limits dependent on the wavelength. This
can be expressed as a condition

$$2\eta\ \delta_b \approx 1. \tag{16}$$

Using Eq.(4) this condition can be understood as implying an optimal
background temperature

$$T_b^{opt} = \frac{h\nu\ /\ \kappa}{\log\ (1 + 2\eta)}\ . \tag{17}$$

For a quantum efficiency of 25% (typical for photodiodes) this becomes
numerically

$$T_b^{opt}(K) = 6\quad 10^{-11}\ \nu\ (Hz)$$

or, alternatively, at a given background temperature there exists an optimal
wavelength for heterodyne detection, given by

$$\lambda^{opt}(\mu m) \quad = \quad \frac{18000}{T_b(K)} \quad . \tag{18}$$

When T_b = 300 K, such as will be the case for observations through the
Earth's atmosphere λ^{opt} = 60 μm; when observing from spacecraft T_b may be
reduced to about 100 K and λ^{opt} increases to about 200 μm. At wavelengths either shorter or longer by a factor α than the optimal wavelength,
the merit ratio increases in first approximation by this same factor α. At
shorter wavelengths this is due to the increased quantum noise (because of
the presence of larger quanta) which affects most the coherent detection
system; at longer wavelengths the increase is caused by the deteriorating
effect of background radiation on the characteristics of the coherent
detector which, in contrast to statements sometimes encountered in the
literature, is not always insensitive to background. At the optimal wave-
length and farther out in the infrared, we have $\delta_b \geq \frac{1}{2\eta} \gg 1$ and the merit
ratio r can in good approximation be expressed directly in the relevant
background radiation temperature

$$r \quad = \quad \frac{2\eta \ T_b + h\nu / \kappa}{\sqrt{2\eta \ T_b \cdot h\nu / \kappa}} \tag{19}$$

from which the above statements follow immediately. Although, therefore,
theoretically heterodyne detection has optimal merits only in a limited wavelength region, around about 100 μm, in practice it may well turn out that for
high resolution spectrographic work, heterodyne detection is the only
workable solution from, say, 20 μm on, since there are no suitable radiation
dispersing elements available - the obvious example is given by all radio-
astronomical detection techniques. However, in the radio region local
oscillator noise is no longer dominating and the formulae become different.
For all other purposes (for instance interferometry), and for observation of
extended sources or continuous spectra, the applicability of heterodyne
detection in far infrared astronomy should not be overestimated.

References

M.W.M. de Graauw, 1975, Thesis Utrecht University.
H.Z.Cummins and H.L. Swinney, 1970, Progress in Optics VIII, 133
                              (North-Holland Amsterdam)

DISCUSSION

Mezger   Could you elaborate on the possible use of incoherent detection in the radio range? It is difficult for me to see how incoherent detection could lead to a better signal-to-noise ratio.

van Bueren   I regret that no analysis has been made yet taking into account the noise properties of the latest solid state detector materials and considering background limiting devices.

PART 5

CONTINUUM EMISSION

FAR INFRARED MEASUREMENTS OF THE ORION NEBULA (M42)

K. Shivanandan, D. P. McNutt, M. Daehler

Space Science Division, Naval Research Laboratory, Washington, D. C. 20375, U.S.A.

and

P. D. Feldman

Department of Physics, Johns Hopkins University, Baltimore, Maryland 21218, U.S.A.

Abstract   Airborne observations of Orion at 100 and 285 μm are reported.

1.   Introduction

The Orion Nebula (M42) Complex has been extensively studied and is proven to be a strong emitter of far infrared radiation. We report the results of measurements of M42 using a 4 arc min beam at 100 μm and 285 μm and compare the results with other observations.

2.   Instrumentation

The apparatus used for the measurements consists of a photometer using photoconductive detectors-gallium doped germanium (Ge:Ga) and n-type epitaxial gallium arsenide (Ga:As) detector. Each of the detector was used in a separate liquid helium cooled dewar with suitable load resistor for optimizing the detector response, filters to define the spectral band and baffles with apertures to define the field of view and reduce the background. The spectral response of the detectors are shown in Fig. 1. A 2 mm circular aperture was used in front of the detectors to define a 4 arc min field of view.

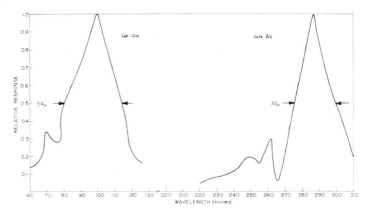

<u>Fig. 1</u>.   The spectral response of Ge:Ga   and   Ga:As   detectors.

3.   Observations

The observations were made with the photometer system mounted on the 31 cm
airborne Lear Jet telescope similar to the one developed by Low, Aumann and
Gillespie (1970). The telescope has a 30 cm diameter ellipsoidal Cervit
primary mirror and a spheroidal silicon secondary mirror of 7.8 cm diameter
with an overall effective focal length of 192 cm. Modulation and background
cancellation was achieved with the secondary mirror oscillating at 95 Hz
and a beam throw of 10$'$ arc.

4.   Calibrations and Results

Absolute and spectral calibrations were made in the laboratory using a black-
body source and a grating spectrometer respectively (Shivanandan, McNutt,
Daehler and Feldman, 1975). The flights were made during the period of
April 10-20, 1974 and observations were carried out at an altitude of 14 km.
In two separate flights Venus and the Orion Nebula (M42) were observed.
Venus was used as a calibration source to compare the detector responsivity
and noise equivalent power to the laboratory measurements. The following
parameters were used:

Diameter of Venus (April 15, 1974): 24.4$''$ arc.

Area of unobstructed telescope: 683 cm^2

Etendue: 7.51 x 10^{-6} cm^{2}sr

Blackbody temperature of Venus: 280°K.

Because of uncertainties in the atmospheric transmission at the observing
altitudes and lack of information on the optical transmission properties
of the telescope at 100 μm and 285 μm, no corrections were made in the
airborne responsivity measurements as presented in Table 1.

The intensity calibrations for the Orion Nebula were based on an effective
blackbody temperature for Venus of 280°K. These results are presented in
Table 2 and compared with other observations. Fig. 2 shows the spectrum of the
Orion Nebula Complex deduced from measurements in the near and far infrared
region.

5. Conclusion

The 100 μm observation agrees with other similar Lear Jet observations and

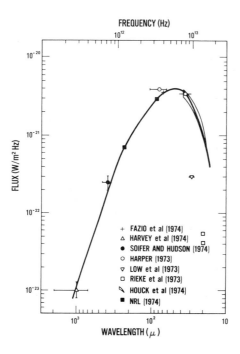

Fig. 2. Results of several studies of Orion Nebula.

Detector	Laboratory		Airborne (Venus)		
	Responsivity (V/W)	NEP (W/Hz$^{\frac{1}{2}}$)	Responsivity (V/W)	NEP (W/Hz$^{\frac{1}{2}}$)	Minimum Flux density (W/m^{2}Hz)
Ge:Ga No. 61 λ_p = 100 μm $\Delta\lambda$ = 34 μm	8.8×10^5	4.5×10^{-13}	2.7×10^5	4×10^{-12}	5.7×10^{-23}
GaAs B6(11-1) λ_p = 285 μm $\Delta\lambda$ = 30 μm	1.3×10^7	4×10^{-14}	4.2×10^6	3×10^{-13}	5.0×10^{-23}

TABLE 1.   COMPARISON OF LABORATORY AND AIRBORNE DETECTOR SENSITIVITIES

Experiment	Bandwidth (μm)	Peak Wavelength λ (μm)	Beam (arc min)	Flux (Wm^{-2}Hz^{-1})	Reference
Photometry	80–115	100	4	3.6×10^{-21}	Present work
	270–300	285	4	7×10^{-22}	
Photometry	65–110	78		4.1×10^{-21}	Harper
	56–500	99	5	3.0×10^{-21}	
	125–500	183		1.4×10^{-21}	
Photometry	50–300	–	8	7×10^{-22}	Low
Grating Spectrometer	80–125 R = 8 μm*	100	5	3.4×10^{-21}	Ward et al
Michelson Interferometer	55–200	100	4	3×10^{-21}	Erickson et al
Fabry-Perot Interferometer	80–135 R ≡ 6 μm*	100	7.4	4×10^{-21}	Brandshaft et al

*R ≡ Resolution

TABLE 2.   OBSERVATIONAL RESULTS OF THE ORION NEBULA (M42) CONTINUUM SOURCE IN THE FAR INFRARED

the 285 μm observation was the first airborne measurements made at this
wavelength. The noise levels observed during flight were ten times those
obtained in the laboratory. Possible sources of this noise were microphonics,
optical offset and telescope guiding errors. The signal-to-noise ratio of
photoconductive detectors improve as the modulation frequency is increased,
hence an improvement in sensitivity could be achieved with higher modulation
frequencies. Microphonics would be reduced by using properly impedance-matched
preamplifiers at 4.2 K.

Acknowledgements

This work was sponsored by the Office of Naval Research. Technical assistance
was provided by A. Lange, R. Tate and E. Wilder. The dewars were provided
by Dr. G. Chanin of Service d'Aeronomie, CNRS, France. The authors wish to
thank W. J. Moore of NRL for providing the Ge:Ga detector, G. Stillman of
the Lincoln Laboratory for providing the Ga:As detector and
Professor Melchiorri of the University of Florence for providing the specially
fabricated narrowband filters used with the Ga:As detectors.
The authors thank the NASA Airborne Science Division, in particular,
Robert M. Cameron and Robert H. Mason, for providing engineering flights on
the Lear Jet and cryogenic support. Their sincere appreciation to the pilots
for making the flights possible.

References

Brandshaft, D., McLaren, R.A. and Werner, M.W., 1975, Astrophys.J., 199, L115.
Erickson, E.F., Swift, C.D., Witteborn, F.C., Mord, A.J., Augason, G.C.,
     Caroff, L.J., Kunz, L.W. and Giver, L.P., 1975, Astrophys.J., 183, 535-
     539.
Fazio, G.G., Kleinmann, D.E., Noyes, R.W., Wright, E.L., Zeilik, M. and
     Low, F.J., 1974, Astrophys.J., 192, L23-L25.
Harper, D.A., 1974, Astrophys.J., 192, 557-571.
Harvey, P.M., Gatley, I., Werner, M.W., Elias, J.H., Evans, N.J., Zuckerman, B.,
     Morris, G., Sato, T. and Litvak, M.M., 1974, Astrophys.J. (Letters), 189,
     L87.
Low, F.J., Aumann, H.H., 1970, Astrophys.J., 162, L79-L85.
Low, F.J., Aumann, H.H. and Gillespie, C.M., 1970, Astronaut.Aeronaut., 8, No.7,
     26.
Low, F.J., Rieke, G.H. and Armstrong, K.R., 1975, Astrophys.J., 183, L105.
Rieke, G.H., Low, F.J. and Kleinmann, D.E., 1973, Astrophys.J., 186, L7.
Shivanandan, K., McNutt, D.P., Daehler, M., Feldman, P.D., 1975, NRL Report,
     7879.
Soifer, B.T. and Hudson, H.S., 1974, Astrophys.J., 191, L83.
Ward, D.B., Harwit, M., 1974, Nature, 252, 27.

DISCUSSION

Welsh (a) What is the precise shape of the secondary mirror modulation at
95 Hz? (b) What is the size and weight of the secondary? (c) Since it is

so light, does it distort when moved at such high frequencies?

Shivanandan a) The shape of the secondary mirror modulation is square wave.
b) Characteristics of secondary mirror used during flight.
  1. Material: Silicon.
  2. Diameter: 7.8 cm.
  3. Radius of curvature: 32.5 cm.
  4. Focal length: 16.25 cm.
  5. Weight: 4.5 ozs.
c) There is some distortion in the wave shape as the frequency is increased, especially at resonant frequencies. This will effect the image quality and the chopper throw. At fixed chopping frequency away from resonance, distortion is minimized.

van Duinen Can you give some figures of merit for your system?

Shivanandan The noise equivalent power for the Ge:Ga detector (λ_p = 100 μm, $\Delta\lambda$ = 34 μm) is 4 x 10^{-13} W/$Hz^{\frac{1}{2}}$ and for the Ga:As detector (λ_p = 285 μm, $\Delta\lambda$ = 30 μm) is 4 x 10^{-14} W/$Hz^{\frac{1}{2}}$ as observed in the laboratory. The sensitivity was 10 times higher onboard the Lear Jet and the minimum flux density using the 30 cm telescope on the Lear Jet was in the range 4000 to 5000 flux units for both the detectors.

Kleinmann What is the NEP you expect to be ahhievable by Ge:Ga and/or Ga:As photodetectors within 4 or 5 years?

Shivanandan It is customary to define the Noise Equivalent Power (NEP) of a detection system with the detector-filter-optics combinations. In a 4.2 K environment we can anticipate Ge:Ga and Ga:As photoconductive detectors to have NEP of the order of 10^{-15} to 10^{-16} W/$(Hz)^{\frac{1}{2}}$ using narrow band filters and cold preamplifiers. Sensitivities of 10^{-14} W/$(Hz)^{\frac{1}{2}}$ have been achieved using these detectors in helium cooled rocket flight experiments. It is too premature for me to forecast what can be done in 4-5 years from now on improving the sensitivities – this is as difficult as forecasting the budget!!!

AN AIRBORNE INFRARED ASTRONOMY PROGRAM: SYSTEM DESCRIPTION AND PRELIMINARY RESULTS

P. Turon, D. Rouan, P. Lena, J. Wijnbergen

and

J.W. Aalders

Observatoire de Paris, 92190 Meudon, France

Abstract The 32-cm airborne telescope has been mounted on the
CV-990 from NASA and extensively used during and after a Spacelab
operations simulation period. The instrument was equipped with a
four-channels photometer covering the spectral range 30-200 μm.
Extensive study of noise behaviour at the tropopause boundary was
made. Preliminary results for M17, W51 and ρOph are given.
The present work is the fruit of a cooperation between the
Space Science Department, University of Groningen and the
Observatoire de Meudon, Groupe Infrarouge Spatial. Contributions
to the instrument are due to the Service des Prototypes du CNRS
and to Dr. Coron.

1. System Description

The airborne astronomy program has been initiated in France in 1970 and a

32-cm telescope has been flown in 1973 and 1974 on a Caravelle aircraft (1).

These flights were first engineering test flights, and then scientific. The

analysis of the data gathered on Sag A and Sag B2 is under progress. This

instrument is versatile enough to be adapted to various aircraft, and this

paper describes an observing campaign on board the NASA CV-990.

The open-port telescope is gyro-stabilized, and gyro drifts are compensated

by an offset star-tracker in a television loop. Tracking stability is of

the order of $15''$ arc rms for a few minutes, and better than $1'$ arc rms

for periods of tens of minutes. The star FOV (3×2^0) is imaged by a

Nocticon tube and tracking is possible on stars $m_v \lesssim 7$. Star field is video

recorded. Offset scanning signals provide mapping capability up to

$30 \times 30'$ arc, by scanning the whole telescope.

The infrared photometer is similar to the one used by the Groningen group in

its balloon flights (2). Equipped with a Low bolometer, it contains four
bands around 35μm, 50μm, 70μm, 140μm, to be commuted sequentially. Cold
diaphragms vary between 0.5 and 6.3' arc. The total weight at the
Cassegrain focus is 10 Kg. The secondary mirror is wobbling, with
amplitudes up to 10' arc. Maximum frequency is 80 Hz, but the system was
operated at 37 Hz. Overall instrument transmission is 14%, assuming an
optical NEP of 3.10^{-14} $W-Hz^{-1/2}$.

The data handling is shared between on-line recording of the video and
analog signals, and digitized maps of the source are both recorded and
visually displayed by a PDP-11 computer. The on-line averaging of signals
provide a quick-look way to analyse data and make decisions.

2. Flight Operations

The system is operated by 2 people. Main problems encountered were due to
aerodynamical dissymetries and induced torques on the telescope, but this
was finally solved by using a proper design spoiler. Part of the mission,
which lasted from June 3 to June 20, 1975, was a Spacelab operations
simulation. Out of the 9 flights, some 12 hours of valuable observing time
were left clear of aircraft, weather, or instrumentation problems. The
height of the tropopause, hence the H_2O content of the overhead atmosphere,
strongly varied during this period of the year at the operating latitudes
($\sim 35^{\circ}$ N).

The system noise of the instrument is to be conservative, NEF $\sim$ 500 $FU-Hz^{-1/2}$,
depending on wavelength and this is essentially not degraded in flight, as
long as the system is not open to the atmosphere. "Sky noise" appears when
actual observing occurs, and is presently the limiting factor of the
instrument. "Sky noise" varies between 1000 and 3000 $FU-Hz^{-1/2}$ and is
strongly dependant on FOV, wavelength, frequency, altitude, air mass, and
atmospheric conditions. Although indications seem to show that it is mostly
due to the inhomogeneous distribution of H_2O and temperature in the
surrounding stratosphere, it should not be excluded, at this stage of data
analysis, that at least part of it might be due to locally induced
turbulence in the airstream along the aircraft. This noise would likely be

eliminated by higher modulation frequency.

3. Results

The results reported here must be considered very preliminary since only
quick-look data have been examined, from 11-20 June flights.

Venus has been available for calibration in some of the flights. Observed
sources are: W 51-M 17 - NGC 7000 area - S-131 - ρ Oph - S Ceph. They have
been observed and mapped in one or several bands.

   W 51 : we have a tendency to find lower values than the one
found by Harvey et al. (3) on the C-141 (by a factor of 2). Spectral
slope is in general agreement with a 80 K blackbody, or steeper, indicating
some emissivity dependance in an optically thin source. The source seems to
be more extended at longer wavelengths, namely to extend over 4 arc min.

   M 17 : a detailed mapping in the 4 wavelengths bands has been
obtained with 1.5 arc min beam.

   ρ Oph : The detection has been achieved with a 6.3' beam
(9' throw), both to increase flux sensitivity and to detect weak gradients.
The detection is positive at a 2.5-4σ level on more than 5 maps, with
flux of the order of 5 KFU in 6.3'. The source seems to be more extended
than the beam size, but this is extremely preliminary. If the molecular
excitation temperature of 30 K is adopted, this flux would correspond to
an optically thin cloud at this wavelength.

   S 131 : the CO cloud observed by Loren (4) in this area has
been mapped. Although weather was bad during this particular part of the
flight, a tentative positive detection of this cloud is reported, at a few
thousand FU level.

There is no obvious indication of detection of the NGC 7000 area or of
S Ceph.

References

(1)   Lena, P. J., 1970, Space Sci.Rev., 11, 131.
(2)   Olthofh. van Duinen, R. J., 1973, Astron.Astrophys., 29, 315.
(3)   Harvey, P. M., Hoffmann, W. F. and Campbell, M. F., 1975, Astrophys.J.,
        196, L 31.
(4)   Loren, R. B., 1975, Astrophys.J., 195, 75.

DISCUSSION

Lemke What are "Spacelab Simulation Flights?" What are you trying to find
out about future Spacelab missions?

Lena ESRO and NASA wanted to find out in these flights answers, or at least
partial answers, on Experimental Operation training, data handling, experiment (there were 6 of them) compatibility in terms of mission profile,
communication with experimenters.

Shivanandan How does the response of the bolometer fall off between 30 Hz
and 80 Hz?

Lena The bolometer has the normal fall-off beyond the 3 dB point which
is, I believe, around 35 Hz.

Shivanandan How does the sky noise vary with altitude?

Lena The position with respect to the tropopause itself is more relevant
than the altitude. But further reduction is needed to answer this question.

Melchiorri With respect to sky noise I want to point out that it should be
correlated at different wavelengths. This means that it is possible to
reduce it by means of two detectors operating simultaneously at two
wavelengths. For example we have found good correlation between ground-level
measurements at 10 μm and 1 mm in the low frequency region (10^{-2} Hz).
Have you data relevant to this question?

Lena We are working with the multibands sequentially, so we have no evidence
on this. But we are considering the possibility of studying the correlation
of noise in two separate apertures, at the same wavelength. If noise is
correlated over an area larger than the source, it could in principle be
eliminated.

Phillips What was the measured value of the temperature of the source
ρ Ophiuchi?

Lena We have no measurements. The first problem was to establish the
existence of 150 μm radiation and we spent all our time mapping in this
single band.

Bussoletti The temperature of 30 K that you infer for dust in ρ Oph is
derived from molecular data and is higher than the expected temperature
because it represents the brightness temperature of the gas not that of the
dust.

Lena The dust temperature does not necessarily have the molecular excitation
temperature of 30 K, but
(a) since we observe in a band 120-190 μm, the temperature cannot be much
less than 10 K, otherwise the Planck function would decrease fast in this
band and decrease the flux,
(b) if one accepts the decoupling between dust and molecules, and a lower
dust temperature, this will mean a larger optical depth to account for the
observed flux. It remains to be seen how far one can push the mass of
dust and the grain emissivity in this particular cloud.

Emerson I do not understand why Bussoletti should assume the dust to be
cooler than the gas. In all far infrared sources that I can think of

(including the KL nebula and ρ Oph) there is an identifiable source of
stellar or protostellar nature in the central part of the far infrared source,
e.g. a compact HII region, heavily reddened cluster of stars or protostars,
B stars etc. In this type of source the dust will be heated by absorption
of photons coming directly or indirectly from the central objects. The gas
is then heated by collisions with the dust grains. In such a case the dust
will be hotter than the gas rather than cooler.

Mezger I agree with Emerson that the cycle - heating of dust grains by
absorption of stellar radiation - heating of gas by collisions with grains -
is the most likely process for heating of dense molecular clouds. Otherwise
it would hard to explain their gas temperatures of about 30 K, i.e. a factor
of 10 above the microwave background temperature.

Andriesse One asks why the infrared emission of HII regions always peaks
around 60 or 70 μm. It is very plausible that the temperature of
insterstellar particles is about 20 K, so in HII regions they have to be
somewhat warmer. If one now assumes that the HII regions are optically
thin and that the emission efficiency increases in some way of other with
frequency, it is readily found that the peak emission should be somewhere
around 60 or 70 μm. The exact position depends little on the temperature,
because the emitted energy depends on a high power of the temperature,
perhaps T^6 or T^7. So there are basic laws, linked to the physical nature
of the dust particles, that fix the observed peaking of emission at the
indicated wavelengths, whereas the particular conditions in the HII region
are only of minor importance.

Greenberg Until we have laboratory measurements of the optical properties of
potentially significant interstellar materials in the far infrared I do not
see how we can make a precise judgement of the temperature of the grains
on the basis of what are still crude spectral energy distributions.

Schultz One has to be careful taking optical constants of bulk material to
explain the properties of very small grains. Skaupy (Phys. ZS, 28, 842, 1927)
has shown that material in the form of a large monocrystal heated in a flame
radiates only at characteristic bands in the infrared, while the same material
powdered radiates in a broad spectral region, with a maximum at the same
wavelength as would be shown by a blackbody of the same temperature.

Greenberg When one grinds up a block of material to a powder, the emissivity
of the ensemble of small particles approaches the blackbody distribution even
though the individual particles have the bulk optical properties. This is
because of the multiple scattering and is true when the optical depth at the
relevant wavelength is large.

Whitworth Surely even without invoking resonances it is not so strange that
the derived grain temperatures occupy a rather small range. The grain
temperature varies as the fifth or sixth root of the heating rate; whilst
the heating rate depends, in the simplest approximation, on the first power
of the exciting luminosity, and on the inverse square of the distance from
the exciting source.

Beckman The answer to the question of optical depth/temperature in ρ Ophiuchi
is best answered experimentally, which is what Lena's experiment has been
designed to do.

TWO STRONG INFRARED SOURCES IDENTIFIED WITH DOUBLE REFLECTION NEBULAE

C. G. Wynn-Williams

Mullard Radio Astronomy Observatory, Cavendish Laboratory.

Cambridge, U.K.

Abstract Two objects from the U.S. Air Force Rocket survey
(Price & Walker, private communication), numbers CRL 2688 and
CRL 618, have each been identified with small double reflection
nebulae illuminated by hidden stars. In one case the star is
deduced to be type F, in the other, type O. The objects may be
protostellar or, alternatively, precursors of planetary nebulae.
These objects have been studied by various astronomers in
different observatories at infrared, optical and radio wavelengths.
Details of CRL 2688 are given by Ney et al. (1975) and Crampton et
al. (1975), while CRL 618 is described by Westbrook et al. (1975).

References

Crampton, D., Cowley, A.P., & Humphreys, R.M., 1975, Astrophys.J. (Letters)
198, L135.

Ney, E.P., Merrill, K.M., Becklin, E.E., Neugebauer, G., & Wynn-Williams, C.G.
1975, Astrophys. J. (Letters), 198, L129.

Westbrook, W.E., Becklin, E.E., Merrill, K.M., Neugebauer, G., Schmidt, M.,
Willner, S.P., Wynn-Williams, C.G., 1975, Astrophys.J. (in press)

DISCUSSION

Swings Does the central blob of CRL 618 have a low or high excitation
spectrum? It appears that peculiar emission-line objects with infrared
excess often show low excitation forbidden lines and/or lines indicative of
high density ($10^6 - 10^7$ electrons per cm^3). D.Allen and I have suggested
that some of these objects, e.g. He 2-446; the "butterfly" M 2-9, could be
young dense planetary nebulae or objects between the Bep stage and the
planetary nebula stage. Therefore knowing the spectrum of the core of the
new CRL source would help in comparing all these peculiar objects.

Wynn-Williams The excitation appears to be fairly low.

van Bueren If your idea of a planetary nebula is correct, do you not find it
strange that this object does not emit any radio radiation? If the estimated
distance is correct, then the radius would be 0.03 pc, and a normal planetary
nebula would be a fairly strong radio emitter.

Wynn-Williams I would not regard this as a "normal" planetary, but as a
precursor of a "normal" planetary. The electron density would be much higher
than normal.

<u>Edmunds</u> Do you have any estimates of the dust mass in these objects? Would the masses involved be any problem from the point of view of production of the dust before the planetary stage?

<u>Wynn-Williams</u> No, but I do not expect there to be a problem on this score.

<u>Gilra</u> For de-reddening the observations of CRL 618, the Whitford curve may not be a good approximation if most of the extinction is local to the source – the local dust may be different from the normal interstellar dust. The observations of CRL 2688 are very interesting, as this is the first object other than a late N star which shows C_3 absorption features. From the observations one notices that the excitation temperature for C_3 may be about 1000 K rather than several thousand, as for an FSI star. At the resolution of Crampton et al. one would not see so much structure in the C_3 spectrum if the temperature were several thousand degrees. That implies it is not a photospheric effect but is some kind of shell phenomenon. The emission features of C_2 (at 5165Å and 5635Å) also imply a low excitation temperature since the line at 4737Å is not seen and those at 5165 and 5635Å originate from the same upper state.

It is worth speculating whether this object is a post-IRC+10^o216 (post carbon star) object. I would suggest that the following observations should be attempted:

(i) Determining the C:O ratio in the stellar atmosphere from oxygen lines in the near infrared and carbon lines at about 4800Å.

(ii) Microwave observations may show the molecular composition of the nebulosity and one may be able to get the $^{12}C^{13}C$ ratio. This may also help in distinguishing whether this is a post carbon star object or a protostellar object.

MILLIMETRE WAVE OBSERVATIONS OF GALACTIC AND EXTRAGALACTIC OBJECTS

P. E. Clegg, P. A. R. Ade and M. Rowan-Robinson

Queen Mary College, London E.1

Abstract   Broad-band observations of galactic and extragalactic
sources at 1 mm are described, and a discussion of the
calibration procedure is given.

1. Introduction

In December, 1974, we continued a series of observations, at a wavelength
around 1 mm, using the 11 m National Radio Astronomy Observatory (NRAO)*
telescope at Kitt Peak, Arizona. Here we report our observed fluxes and
upper limits for various galactic and extra-galactic sources. Little will
be said here in interpretation of these data; this will be done in a
subsequent paper. The calibration of the observations will, however, be
discussed in some detail as we have evidence of large variations in atmospheric attenuation which can lead to considerable errors in the assignment of
fluxes. Indeed, it is our detailed investigation of this point which has
delayed interpretation of the data. This discussion will be followed by
presentation of fluxes and a map of Galactic sources, flux limits on galaxies
within 3 Mpc and measured fluxes from Seyfert and N-galaxies and QSS's.

2. Calibration

The method of calibration has been described in detail in Rowan-Robinson et al
(1975), referred to here as Paper I. Frequent observations are made of the
planets Jupiter (assigned a brightness temperature of 150 K, Saturn (194 K),
Mars (225 K) and Venus (291 K). These observations we used to calibrate flux

---

*NRAO is operated by Associated Universities, Inc., under contract with the
National Science Foundation.

measurements via the equation, valid for a plane stratified atmosphere,

$$V = k_2 W_o \exp - a(t) \sec z \qquad (1)$$

where V is the measured voltage, W_o is the flux from the source incident on the upper atmosphere, z is the zenith distance, $a(t)$ the (time dependent) atmospheric attenuation and k_2 is the telescope-receiver conversion factor between flux and voltage. For a calibrating planet, k_2 and $a(t)$ are unknown; were a constant, k_2 and a could be obtained simply by plotting ln V against sec z. The large variability observed for a , however, precludes this. Instead, we choose that value of k_2 which minimises the correlation between the quantity

$$\frac{\ln k_2 - \ln V/W_o}{\sec z}$$

and sec z over the whole run. The selected value of k_2 is then used in Eq.(1) to obtain $a(t)$ for any given observation. Finally, the values of $a(t)$ derived from all planetary observations are fitted with a smooth curve which is used to interpolate a as a function of time; the result for December, based on 109 observations, is shown in Fig.1. The very large diurnal variation in a is immediately obvious and it is clearly important to establish that the observed variability of a is real and not a systematic effect.

The reality of the effect is supported by readings taken with a water vapour meter which uses the sun to measure the attenuation in the near infrared. From this attenuation, the vertical precipitable water vapour w is inferred and we have used the empirical relation

$$a = 0.138 \ w \ (mm)$$

to fit these observations with the planetary calibrations in Fig.1. It can be seen that the agreement between the variation of the two independently derived values of a is good.

We have also considered the possibility of variations in telescope and receiver gain and of the breakdown of the plane-parallel atmospheric approximation.

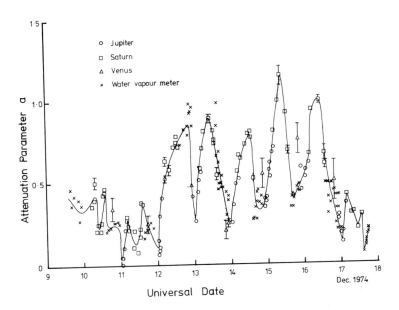

<u>Fig. 1</u> Attenuation parameter, a(t) plotted as a function of
Universal date; the smooth curve was used to obtain the
attenuation for any given source.

(a) <u>Telescope gain</u> We have found no correlation between a(t) and ambient
temperature, telescope beamwidth or focus setting, all of which may be
expected to be measures of telescope gain.

(b) <u>Receiver gain</u> The only section of the receiver liable to gain variation
is the front end. Variations in gain of this section should affect equally
the signal and the noise. (The value of $\sigma^2 t_{int}$, where σ is the variance in
a particular observation with integration time t_{int}, is completely
uncorrelated with sec z and inferred water vapour, suggesting that we have
no residual sky noise.) We find, however, that the noise was practically
constant during the whole run. We therefore believe that the gain of the
receiver is constant, and this belief is supported by laboratory estimates
of possible gain variation.

(c) <u>Plane-stratified approximation</u> The validity of this approximation appears
to be confirmed by (i) the continuity of the a(t) curve obtained while
Jupiter was going from transit to setting (0-5 hrs U.T.) with that obtained
while Saturn was going from rising to transit (4-9 hrs U.T.), and (ii)
agreement between planetary and water vapour meter attenuations.

TABLE 1. GALACTIC SOURCE FLUXES.

Source		R.A. (1950)	Dec. (1950)	Flux (1 arcmin FWHP Beam)
Orion	BN	$05^h32^m46^s.7$	$- 05°24'17''$	128 ± 7.7 Jy
	KL	05 32 46.8	$- 05\ 24\ 28$	170 ± 23.6
	A	05 32 49.7	$- 05\ 25\ 12$	47.8 ± 14.1
	OMC2	05 32 59.1	$- 05\ 12\ 10$	18.4 ± 6.5
W49		19 07 49.9	09 01 18	55.5 ± 11.9
W51		19 21 24.3	14 24 42	125.8 ± 11.9
W3		02 21 57.0	61 52 48	41.7 ± 8.8
		02 21 50.5	61 52 21	21.5 ± 6.6
		02 21 53.0	61 52 20	40.0 ± 24.5
DR21		20 37 14.2	42 09 07	21.6 ± 5.0
		20 37 14.2	42 12 00	21.1 ± 5.1
W58	K3-50AB	19 59 50.2	33 24 20	24.0 ± 8.1
Crab	Pulsar	05 31 31.5	21 58 55	31.2 ± 11.0
	Tau A	05 31 30.0	21 59 43	73.3 ± 27.9
Sag	B2	17 44 10.7	$- 28\ 22\ 04$	124.0 ± 35.9
	1' N	17 44 10.7	$- 28\ 21\ 04$	149.4 ± 28.4

In view of these considerations we are convinced that the inferred
variation in attenuation is real and that continual monitoring of planetary
fluxes is therefore essential for accurate flux calibration.

We have used the a(t) curve given in Fig.1 to reduce the fluxes discussed
below.

3. Fluxes of Galactic Sources

Table 1 gives the fluxes observed, within a 1' arc beam, observed from
various positions in galactic sources. The nominal effective frequency $\tilde{\nu}$
of these observations is   243 GHz ($\equiv$ 1.23 mm). $\tilde{\nu}$ is defined by

$$\tilde{\nu} = \int \nu\ S(\nu) f(\nu) d\nu / \int S(\nu) f(\nu) d\nu$$

$$(2)$$

where $S(\nu)$ is the spectrum of the source and $f(\nu)$ the filtering produced
by the atmosphere, telescope surface and internal filters. We have assumed

$$S(\nu) \alpha \ \nu^{\alpha}$$

with $\alpha = 3$ for these Galactic sources.

We have integrated the fluxes from Orion and Sag B2. For Orion, the total
flux within 3' arc of BN is 1613 ± 484 Jy. A crude estimate of the
millimetre luminosity is then given by $4\pi d^2 \ \tilde{\nu} \ S(\tilde{\nu})$ and is $(3 \pm 1) \times 10^1 L_{\odot}$
assuming a distance d of 0.5 kpc.   For Sag B2, we have assumed that the
brightness distribution is Gaussian of $2'.4 \pm 0'.4$ FWHP centred 30" N of Sag
B2; this gives a flux of 470 ± 160 Jy and a luminosity of $(3.6 \pm 1.2) \times 10^3 L_{\odot}$
for a distance of 10 kpc. Of course, this luminosity can only be an extreme
lower limit, since we expect the source to be very much larger than this.
Nevertheless, it is a useful figure to bear in mind when discussing extragalactic sources.

We have also obtained a 50×17 point grid map of Orion at 20" arc interval,
consisting of 12,160 datum points and a total integration time of about 6
hours. This map is the most detailed yet obtained for Orion at 1 mm; it
will be published subsequently when the agreement in its overall features
with that of CO 2 - 1 maps obtained by Phillips et al (1975, to be
published) will be discussed.

4. Extragalactic Flux Limits

In Paper 1 we published upper limits on the flux from extragalactic sources,
representing a wide range of astronomical objects. We have now obtained
improved limits (due to better balancing of the beams in our switched-beam
system) on selected sources, including galaxies within 3 Mpc. Figure 2 shows
the value of $L_{\nu}/4\pi$ for these galaxies, plotted against equivalent redshift,
assuming $H_o = 50$ km s^{-1}Mpc^{-1} and using

$$\log \frac{L_{\nu}}{4\pi} = 2 \log \ z \ + \ 2 \log \frac{c}{H_o} \ + \ \log S(\nu) \qquad (3)$$

The upper line on the curve corresponds to a flux density limit of 25.5 Jy,

the lower line to 6 Jy. For the upper curve, we would expect to see Sag B2,
using the given luminosity, out to log z = -5.15 corresponding to 43 kpc.
Clearly, this is not a very strong limit, however, it should be borne in mind
that the given luminosity is very much a lower limit, probably by a factor of
∿100. If this be so, we should detect Sag B2 and to ∿0.5 Mpc so that we
begin to have useful limits on millimetre luminosities of galaxies. We
believe it possible to reduce the limiting flux densities by about an order of
magnitude by detector improvements.

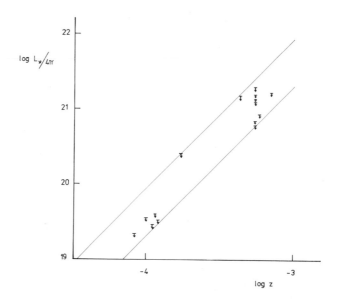

Fig. 2 Upper limits on millimetre luminosities of galaxies
within 3 Mpc: log L$_\nu$/4π (W sr^{-1} Hz^{-1}) - log z.

5. Measured Extragalactic Fluxes

Table 2 gives fluxes measured from two Seyfert galaxies, an N-galaxy and two
QSS's at an effective wavelength of 1.67 mm, in December 1974. Also shown are
the results obtained in February 1974, with and without a high-pass filter.
In all cases the effective wavelength is obtained from Eq.2 with $\alpha = 0$. The
spectra of these sources, from radio to optical wavelengths is shown in
Figures 3a and b.

Inspection of Table 2 will show that the results for both 3C84 and 3C273 are
consistent with a steady flux level and, from Fig.3, with the extrapolation
of the radio synchrotron spectrum.   The 1.57 mm fluxes previously reported
for NGC4051 and 3C147 are thrown into doubt by failure to detect them on
either occasion without the filter. In both cases, the upper limits are
consistent with the extrapolated radio spectrum. On the other hand, 3C120
shows a decrease from a 4σ detection at 15-20 Jy to a 2σ upper limit of
4 Jy.   We believe that this represents genuine long-term variability of
3C120. We are currently monitoring, with E.E.Epstein, J.D.G. and E.Rather,
the short-term variability of several sources.

TABLE 2. EXTRA GALACTIC SOURCE FLUXES

		Flux (Jy)		
Source	Type	$\tilde{\nu}$ = 1.67 mm Dec.1974	Feb.1974	$\tilde{\nu}$ = 1.57 mm Feb.1974
NGC1275 = 3C84	Seyfert	21.6 ± 1.2	14 ± 8.6	19 ± 3.5
NGC4051	Seyfert	≤ 4.8	≤ 46	16 ± 5.3
3C120	N-galaxy	3.0 ± 2.0	15 ± 3.8	20 ± 6.3
3C147	QSS	≤ 10.5	≤ 7	23 ± 8.5
3C273	QSS	12.4 ± 1.7	52	14 ± 2.9

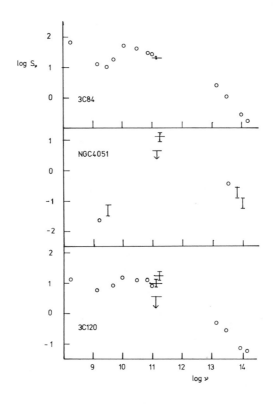

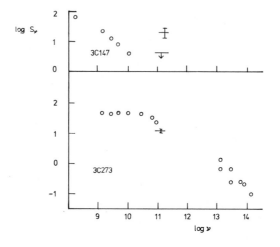

Fig. 3a,b Spectra of extragalactic sources. The units of S_ν are Jy and of ν are Hz.

Acknowledgments

It is a pleasure to thank the National Radio Astronomy Observatory for observing time and the staff of NRAO, Tucson division for tireless and expert assistance. We thank the Science Research Council for financial support for this work; P.A.R. Ade acknowledges tenure of an S.R.C. Fellowship.

Reference

Rowan-Robinson, M., Clegg, P.E., and Ade, P.A.R., Millimetre emission from extragalactic objects - I, Mon.Not.R.ast.Soc. (1975, in press)

INTERPRETATION OF FAR INFRARED EMISSION

J. P. Emerson

and

R. E. Jennings

Department of Physics and Astronomy,

University College, London W.C.1

Abstract Far infrared sources of radiation are interpreted in
terms of emission from dust clouds heated by hot stars, and the
problems of the location, temperature, composition and amount of
dust are discussed. The dust appears to be present both inside
and outside the HII region, and grains (or grain mantles) of ice
fit the data well.

Introduction

In the last few years a large number of sources of far infrared (FIR) radiation
have been detected from balloon and aircraft altitudes. The Infrared Group
at University College London (UCL) has been involved in a balloon borne
program of broad-band photometry in the 40-350 μm band. Here some problems
in the interpretation of this kind of data are considered and some possible
ways of solving these problems suggested.

The FIR radiation is produced by thermal emission from dust grains at
temperatures of several tens of Kelvins, the temperature of the grains being
maintained against the energy loss in the form of FIR emission by absorption
of photons incident on the dust at shorter wavelengths. As a consequence of
this the amount of energy available to power the FIR emission is limited by
the total luminosity of the objects within or close to the dust cloud. Thus
the most luminous FIR sources are found in the vicinity of HII regions,
which contain numbers of the hottest and most luminous O stars, whilst
around reflection nebulae produced by early B stars the FIR emission is
much weaker due to the considerable reduction in energy output of a B star

compared with an O star.

Although we have the general picture given above the details are much less
clear and in particular there are four questions, all somewhat interrelated,
which need to be answered for a fuller understanding of the processes taking
place.

(i)   LOCATION? Where is the dust?

(ii)   TEMPERATURE? How hot is the dust?

(iii)   COMPOSITION? What is the dust made of?

(iv)   AMOUNT? How much dust is there?

Location?

Figure 1 is helpful in giving an idea of the problem of the location of the
dust. The plot shows the total FIR luminosity for 37 sources associated
with HII regions which have been measured by UCL (the measured luminosities
have been increased by a factor of 1.48 to take account of the flux falling
outside the 40-350 μm band for an 80 K blackbody, but this does not
greatly change the look of the plot), plotted against the number of Lyman
continuum photons/sec required to maintain the ionisation, as derived from
radio continuum observations. As most of the radiated energy of the underlying
stellar or protostellar objects is probably ultimately absorbed by dust the
FIR luminosity should be a good approximation to the total luminosity of
the objects in the region. Also plotted is the relationship between these
quantities for Zero age main sequence (ZAMS) stars (and an OB cluster to
cover the case of multiple stars). As most stars are on the main sequence
this line should be appropriate to most of the objects. The important thing
to note is that the observational points lie above and to the left of the
ZAMS-OB cluster line. There are three possible interpretations of this plot.

(1) The sources are deficient in Lyman continuum photons appropriate to their
luminosities by an amount proportional to their distances to the left of the
ZAMS-OB cluster line. This would suggest that dust particles absorbed some
of the Lyman continuum photons directly and hence that there is a significant
amount of dust inside the HII regions. This has important consequences for
the ionisation structure of HII regions.

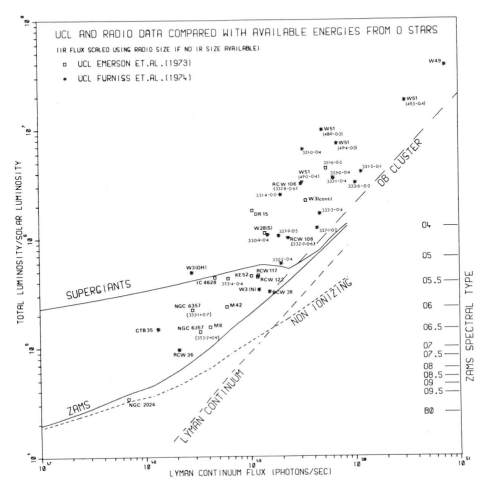

Fig. 1. Far infrared luminosities for 37 sources plotted against the number of Lyman continuum photons/sec required to maintain the ionization.

(2) The sources have excess luminosity over that appropriate to stars emitting that many Lyman continuum photons/sec. This would suggest that there are additional sources of energy to the ionising stars, by an amount proportional to the distance that the points lie above the ZAMS-OB cluster line.

(3) The model atmospheres from which the ZAMS line is constructed are incorrect and/or the relative number of stars of different spectral types as given by the Salpeter initial luminosity function are incorrect.

It seems unlikely that (3) is the case as other (non FIR) lines of evidence
do not suggest any gross discrepancies between predictions and the observations.
Possibility (2) also seems unlikely as the major cause of the discrepancy as
this would require all the objects observed to be special in having an
anomalously large number of highly luminous late type (i.e. cool) stars in
them. The most natural explanation seems to be (1) and requires that the
optical depth for absorption of Lyman continuum photons by dust be typically
about 1. Note that although (1) is favoured this only shows that dust
inside the HII region is absorbing Lyman continuum photons (and hence
probably also radiating the energy away in the FIR), but, as we shall see,
there may still be a contribution to the emission from dust outside the HII
region.

Further evidence for the presence of significant amounts of dust in HII
regions is provided by the correlation between the apparent abundance of
helium and the infrared excess (Mezger, Smith & Churchwell 1974, Emerson and
Jennings 1976) which also indicates that the dust has different absorption
cross-sections for helium and hydrogen ionising photons.

Comparison of FIR and radio maps of HII regions generally indicates a
correspondence between the FIR and radio contours (e.g. compare the map
of W3 in Fig. 2 (from Furniss, Jennings & Moorwood 1975 (FJM)) with radio
maps at similar resolution such as that by Schraml & Mezger 1969) also
suggesting significant emission from dust inside the HII region. Note
however that at the resolutions used it is difficult to distinguish between
emission coming predominantly from inside the HII region, and from a shell
around it.

Thus although the above considerations definitely indicate that dust absorbs
Lyman continuum photons in HII regions it is not clear whether the remaining (mainly non-ionising) stellar radiation is absorbed largely in the HII
region or in the surrounding molecular clouds or at the boundary between them.
Going on from the energetics and morphology discussed so far to attempt to
interpret the emission in terms of temperature, composition and amount of
dust can throw further light on these problems.

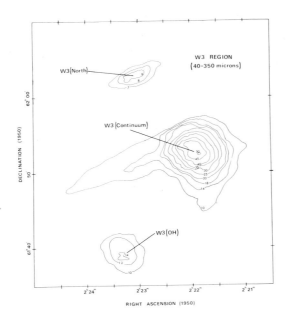

Fig. 2.   Far infrared map of the W3 region.

Temperature? Composition? Amount?

The infrared luminosity between λ_1 and λ_2 µm produced by n_d grains
cm^{-3} of radius a, absorption efficiency $Q(\lambda)$ and temperature T_d is

$$L_{\lambda_1-\lambda_2} = n_d \int_{\lambda_1}^{\lambda_2} 4\pi a^2 Q(\lambda)\pi B(\lambda_1 \ T_d)d\lambda$$

where $B(\lambda, T_d)$ is the Planck function and the emission is optically thin
(as found observationally). The mass of dust is

$$M_d = \frac{4}{3}\pi a^3 n_d \rho$$

where ρ is the solid dust density (1 gm cm^{-3}). Combining the two equations
we have

$$M_d = \frac{\rho\, L_{\lambda_1 - \lambda_2}}{3 \int_{\lambda_1}^{\lambda_2} \frac{Q(\lambda)}{a} \pi B(\lambda, T_d)\, d\lambda}$$

Now in the limit $2\pi a/\lambda \ll 1$, as is appropriate in the FIR for grains of normal size, $Q(\lambda) \propto a/\lambda^b$ where b is typically about 2, so that the integral term is independent of a but proportional to a power of T_d which can be high e.g. if b = 2 and $\lambda_1 = 0$, $\lambda_2 = \infty$ the integral is proportional to T_d^6. The UCL measurements have been made with λ_1 = 40 µm and λ_2 = 350 µm and for this case the lower the value of T_d the stronger the dependence of the integral on T_d. For the ice and silicate grains considered below for W3 the integral depends on the 2nd to 3rd power of T_d in the temperature range discussed. The strong dependence of M_d on T_d means that in general T_d must be quite well determined to give an accurate value for M_d. Unfortunately in the case of a broad-band measurement (as made by UCL) no information on the dust temperature is obtained from the measurement. (With this in mind, on the next set of flights a 3 band photometer will be used to try to get some temperature information.)

This sensitivity of the dust mass to temperature can be turned to advantage if an estimate of the dust mass can be obtained by other means (e.g. from gas mass and gas to dust ratio), and then used to find at what temperature a certain mass of dust, of a given type $(Q(\lambda))$, must be to give the observed FIR luminosity. Comparison of the required temperatures of different grains in different places (e.g. M_d in an HII region differs from M_d in a neutral shell) with other temperature indicators can then provide a basis for deciding on the true physical situation.

To illustrate the method two components (W3 Con and W3 OH) of W3 (see Fig. 2) have been chosen. The discussion is intended to be illustrative rather than exhaustive and a fuller discussion of this type of approach for more sources will be given in Emerson (1976).

Rearranging the last equation and expressing L_{40-350} and M_H in solar units we have

$$\int_{40}^{350} \frac{Q(\lambda)}{a} \, \pi B(\lambda_1 \, T_d) d\lambda \;\; = \;\; 0.64 \left(\frac{M_H}{M_d}\right) \frac{L_{40-350}}{M_H}$$

where (M_H/M_d) is the gas to dust ratio of mass density multiplied by the solid grain density, and is assigned the currently popular value of 200. The integral has been evaluated as a function of T_d and is tabulated in Emerson (1976) for silicate and ice grains using absorption efficiencies (Q) from Knacke & Thomson (1973) and Irvine & Pollack (1968) respectively. The results for ice grains may be expected to hold quite well for ice mantles with refractory cores, especially if the cores are small, as the mantle dominates the optical properties.

Two extreme cases for the location of the dust are considered, to illustrate the possibilities. The 40–350 μm luminosities used are from FJM (1975).

1st assumption: All dust is in the HII region.
Taking the results of Schraml & Mezger (1969) for M_H and increasing them by a factor of $(3.1/2.6)^{2.5}$ to take account of the distance now assumed (3.1 instead of 2.6 kpc) we find for W3 OH, $M_H < 47 \, M_0$ and for W3 OH $M_H < 8 \, M_\odot$, so that using the equation and the tabulations of the integral we find for ice grains $T_d > 141$ K for W3 Con, and $T_d > 158$ K for W3 OH, whilst for silicate grains the corresponding temperatures are > 252 K and > 293 K respectively. Strictly speaking this might be considered to rule out pure ice as it evaporates at 100 K in the environment of an HII region. However in reality the grains will not be pure and a higher evaporation temperature might be appropriate for such 'waxy' grains, which might have similar optical properties to ice.

2nd assumption: All dust is in the neutral shell surrounding the HII region
We can estimate M_H via M_{co} using the number column densities of CO given by Wilson et. al. (1974), the mass of a CO molecule, a distance of 3.1 kpc and the measured FIR sizes of 2' arc for W3 Con (FJM 1975) and $< 1.2'$ arc for W3 OH (Fazio et. al. 1975) to find for W3 Con, $M_{co} = 2.28 \, M_\odot$, and for W3 OH, $M_{co} < 0.38 \, M_\odot$. The amount of CO relative to H is limited by the relative cosmic abundances of C and H so that the mass density ratio of hydrogen to CO must be ≥ 109 (the equality holds

if all C is in CO). Thus for W3 Con, $M_H \geq 248 M_\odot$, and for W3 OH,
$M_H \simeq 41 M_\odot$ (assuming the $\geq$ and < cancel out), so that we find for ice
grains that $T_d \leq 80$ K in W3 Con and $T_d \simeq 87$ K in W3 OH whilst for
silicates we have ≤ 118 K and $\simeq 131$ K respectively.

Discussion: In the case of W3 Con, Harper (1974) has made multiband
observations and finds a colour temperature of 77 K. Looking through the
various temperatures calculated above we see that icy dust outside the HII
region (especially if the equality holds) seems to explain the observed
emission best. The dust in W3 OH seems to need to be hotter than that in
W3 Con and a similar calculation for dust in the HII region of W3 North
indicates it must be rather cooler than the other two sources. This
temperature range may indicate an evolutionary sequence with W3 OH being
the youngest and W3 North the oldest of the sources.

Bearing in mind the illustrative nature of this discussion a strong conclusion
is not warranted, but if the results are accepted at face value they do show
that dust in the neutral shell surrounding the HII region is an important
source of emission and that ice grains (or grains with ice mantles) fit the
emission quite well. However Fig. 1, as discussed above, provides compelling
evidence for dust inside the HII region. Determination of the relative
importance of emission from dust interior and exterior to the HII region
must await further observational evidence in the form of higher spatial and
spectral resolution studies, as well as more information on the emissivity
of different kinds of dust in the FIR.

A Spectrum of W51

As a step in the direction of higher wavelength resolution a Michelson
interferometer (similar to that described by Furniss et al, p.71, this
volume) was used on the UCL platform to obtain the spectrum of W51 shown
in Fig. 3, which is taken from Alvarez (1974) (see also Alvarez et. al. 1975).
Alvarez was not able to find a material which could reproduce the observed
spectrum, and although ice has features in this wavelength region they do not
correspond to the features observed, which could possibly be indicative of
the 'dirtiness' of the ice. Spectra such as this, if confirmed by further
measurements, will be of great value in tying down the dust composition and

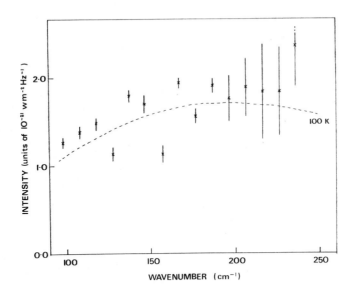

Fig. 3.   45-100 μm   spectrum of   W51.

temperature.

NGC 2023, a Reflection Nebula with   FIR   Emission

A recent paper by Emerson, Furniss and Jennings (1975) also provides evidence
for ice (or ice mantle) grains being important in FIR emission, and suggests
that in dense clouds the gas and dust temperatures are coupled. For a full
discussion the original paper should be consulted, but for completeness a
brief summary is presented here.

Most FIR radiation detected to date has been observed towards HII regions,
but the composition of the grains is not known and it is not clear if the
grains most responsible for the observed emission lie inside the HII regions
or in the molecular clouds which surround them. To avoid these uncertainties
we observed radiation, in the 40-350 μm band, from a reflection nebula and
molecular cloud NGC 2023. This source is not associated with a prominent
HII region.

For a given grain composition and temperature we can calculate the density
of dust required to produce the observed flux within the measure infrared
size of the object. This density is independent of the grain size but very
strongly dependent (T_d^4 to T_d^6) on the grain temperature. Radio molecular
line observations of CO give values for the gas density and more directly
the gas temperature (43 K). We assumed that the gas temperature is proportional to the dust temperature in dense clouds like NGC 2023 and calculated
the gas to dust ratios for ice, silicate and graphite grains for NGC 2023
and two other molecular clouds (KL and OMC2) which have different gas
temperatures (75 and 53 K).

For any given grain type we found that the gas to dust ratios are similar
for all three objects. Because of the strong temperature dependence this
suggests that our assumption that the gas and dust temperatures are
proportional is valid. The gas to dust ratios for the three grain types
are, however, quite different so we can use the results to try to determine
the grain composition. If the grains are made of ice, and the dust and gas
temperatures are equal, the gas to dust ratios in all three objects are close
to the value in the interstellar medium. This result ties in with evidence
from other wavelength regions which suggests that ice grains are more
important than silicate or graphite in the interstellar medium.

A further conclusion results from analysis of an ultraviolet (1350-2900 $\overset{o}{A}$)
spectrum (from the S2/68 telescope on the TD1 satellite) of the exciting
star of the reflection nebula. Using this spectrum it can be shown that the
energy removed by dust from the stellar ultraviolet radiation field is
sufficient to power the observed far infrared emission. The gas is then
presumably heated by collisions with the resulting warm dust grains.

The observability of the effects of the dust grains from the ultraviolet
through the optical to the infrared and the ability of CO and other
molecular lines to probe the gas in the cloud make this object very
important for determining the properties of dust grains and their coupling
to the gas, and indicates the importance of making far infrared observations
of other similar objects.

References

Alvarez, J.A., 1974, Ph.D. thesis, University of London.
Alvarez, J.A., Furniss, I., Jennings, R.E., King, K.J. and Moorwood, A.F.M.,
     1975, in 'HII regions and the Galactic Centre' ed. A.F.M. Moorwood,
     ESRO SP-105.
Emerson, J.P., 1976, Ph.D. thesis, University of London, to be submitted.
Emerson, J.P. and Jennings, R.E., 1976, in preparation.
Emerson, J.P., Furniss, I. and Jennings, R.E., 1975, Mon.Not.R.astr.Soc.,
     172, 411.
Fazio, G.G., Kleinmann, D.E., Noyes, R.W., Wright, E.L., Zeilik, M. and
     Low, F.J., 1975, Astrophys.J., 199, L177.
Furniss, I., Jennings, R.E. and Moorwood, A.F.M., 1975, Astrophys.J., in
     press.
Harper, D.A., 1974, Astrophys.J., 192, 557.
Irvine, W.M. and Pollack, J.B., 1968, Icarus, 8, 324.
Knacke, R.F. and Thomson, R.K., 1973, Publ.Astr.Soc.Pacific, 85, 341.
Mezger, R.G., Smith, L.F. and Churchwell, E., 1974, Astr.Astrophys., 32, 269.
Schraml, J. and Mezger, P.G., 1969, Astrophys.J., 156, 269.
Wilson, W.J., Schwartz, P.R., Epstein, E.E., Johnson, W.A., Etcheverry, R.D.,
     Mori, T.T., Berry, G.G. and Dyson, H.B., 1974, Astrophys.J., 191, 357.

DISCUSSION

Gilra What is the value of E_{B-V} for HD37903?

Emerson 0.36 mag.

Gilra Did you obtain the ultraviolet extinction curve for HD37903 before subtracting the "mean" interstellar extinction curve? We know that the ultraviolet extinction curve is different for different stars.

Emerson The amount of extinction due to the NGC 2023 dust cloud was calculated in two ways, one using a mean interstellar extinction curve and the other by comparison with an unreddened model B1.5V star. The two methods gave quite similar results.

van Duinen Instead of going through these quite complicated arguments through CO to estimate grain temperatures, I suggest you use the results of far infrared multicolour photometry (Olthof & van Duinen, Astron.Astrophys. 29, 315, 1973, Olthof, Astron.Astrophys. 33, 471, 1974). And in addition to your interpretation of the upward deviation from the ZAMS curve in the infrared-radio plot, I suggest you look into the proposal of Olthof (Ph.D. thesis, University of Groningen, 1975) invoking absorption of $\lambda > 912$ Å photons by dust as the required additional energy source.

Emerson Certainly in the cases where multiband observations are available these give a more direct value for the temperature. My intention was to show how a broad-band measurement can be interpreted in terms of different temperatures under different assumed conditions. To distinguish between these different conditions, the temperature must be determined from observations, as in the core of W3. On your second point, taking the Salpeter luminosity function and recent model atmospheres there is not sufficient energy to explain the data in the way you suggest. For any individual object one might make a case, but it is hard to believe that all the sources, including the giant HII regions, could be explained in this way. From optical observations it is known that there is dust in HII regions and it seems very reasonable that it should absorb Lyman continuum photons. The average dust optical depth to the Lyman continuum ratio is only of the order of 1.

Wannier I have made maps of NGC 2023 at the J = 1 - 0 transition of
CO and ^{13}CO. Although the CO line intensity peaks very sharply on
NGC 2023, the ^{13}CO line intensity shows a very much smaller relative peak.
I therefore suggest that the CO antenna temperature results from a thin
layer of quite hot CO. This result might increase your estimate of the
real kinetic temperature of CO and thus affect your results.

Emerson I did assume that the CO emission was optically thick so that the
CO brightness temperature represented the gas kinetic temperature. Nevertheless the results suggest that the dust and CO brightness temperature are
proportional, regardless of their absolute values.

Rowan-Robinson How can the grains all be at the same temperature?

Emerson The grains were assumed to be all at a single temperature because I
do not feel that the available data yet warrants a more complex model.

Aannestad I think one should be very careful about deducing the location of
the dust assuming a constant temperature over the source. If a temperature
variation is taken into account, one may show that with ice grains most of
the dust has a temperature in the range 50 - 100 K even within an HII
region like W3A.

Emerson Yes. However from the observational point of view, it is not clear
what form such a temperature variation might take.

Greenberg H_2O ice is known to be only a small portion of the grain mantle.
The probably complex real mantle material may have significantly different
far infrared absorptivity from H_2O ice. The composition of the dust is
probably a mixture of core-mantle particles in the 0.1 μm size range and a
large number of much smaller bare particles (probably silicates) in the
0.005 μm size range. Each of these contributes differently to the infrared
emission (each component has a different temperature and different complex
refractive index) and this changes the effective radiation temperature.

Emerson Water ice is the only volatile component of mantles for which the
necessary optical constants are available in the far infrared. In addition
water ice is the only mantle component identified spectroscopically (at
3.1 μm). Thus it is not possible to do more realistic grain mantles at
present. Also it does not seem to me that the data which the model must fit
are sufficiently extensive to justify the introduction of a bimodal grain
distribution.

L.F. Smith P.G. Mezger and I have used the improved diagram of He^+/H^+ vs.
IR excess given by Jennings (1975) to re-derive the absorption cross sections
of the dust in the HII regions for the wavelength ranges $\lambda912-504$ Å and
$504-228$ Å in the way first applied by Mezger, Smith and Churchwell (1974).
We set $a_0 = x\sigma_{He}/x\sigma_H$, as before, but now $b = \tau_{\lambda>912}/\tau_H$ and $\tau_{\lambda>912}$ includes
any absorption outside the HII region, including absorption of secondary
photons from the HII region ($\lambda > 912$ Å). It is safe to assume that within
the HII region $\tau_{\lambda>912} \geq \tau_H$, since τ_H is very low. Hence $b \geq 1$, and
varying b from 1 to ∞ corresponds to the full range of possibilities:
from no absorption outside the HII region to absorption of all stellar
radiation $\lambda > 912$ Å and all secondary photons. The effect of large b is
to increase the IR excess (especially when τ_H is small) but not to affect
the He^+/H^+ ratio. The observed points fall between ($a_0 = 7$, $b = 1$) and
($a_0 = 5$, $b = ∞$); thus $a_0 = 6 \pm 1$ may be firmly set.
The values of $x\sigma_{He}$ in the solar neighbourhood range from 0.7 to 2.0 $x\sigma_V$
(Smith, 1975). The corresponding range of $x\sigma_H$ is 0.1 to 0.4 $x\sigma_V$, which
is below the theoretical lower limit 0.7 $x\sigma_V$ for the diffuse interstellar
matter in the solar neighbourhood (see Smith 1975). The contradiction could
be resolved if the number of large grains per H-atom in the HII regions was
less than the diffuse interstellar medium in the solar neighbourhood by a
factor of about 3. (1975, "HII regions & related topics", ed.T.L. Wilson, D.
Downes).

THE DISTRIBUTION OF IONIZED GAS AND DUST IN W 3(A) AND W 3(OH)

P. G. Mezger

Max-Planck-Institut für Radioastronomie, Bonn, W. Germany

Abstract Recent radio and infrared observations are summarized and interpreted. A model fit yields the distribution of dust and ionized gas. The models are compared to a possible evolutionary sequence of O-stars and associated shell of gas and dust. The problem of dust inside HII regions and its effect on the ionization structure is discussed.

1. Introduction

W 3 is the youngest in a sequence of four giant HII regions, which are located in the Perseus arm at a distance of about 3 kpc from the sun. While three of these giant HII regions, viz. IC 1848(=W 5), IC 1805(=W 4) and IC 1795 are observable in the optical range, W 3 is nearly completely obscured by dust and therefore is best observed in the radio and IR range. Due to its proximity to the sun and its high declination, it is probably the best studied giant HII region in our Galaxy. The following review is a summary of three papers (Mezger and Wink, 1975a, b; Krügel and Mezger, 1975; for detailed references to earlier work, we refer to these papers). Here, we concentrate on three points related to far IR-observations:

i)   The astrophysical information which can be obtained by combination of radio and far IR-observations.

ii)   The relevance of this information for the early evolutionary stages of O-stars and compact HII regions.

iii) The effect of dust on the ionization structure of an HII region.

2. Radio and IR-observations

In the radio range we observe the free-free continuum and the bound-bound
recombination line emission.   In the IR at wavelengths > 10μm, we observe
the thermal radiation from dust grains which are heated by absorption of
stellar or plasma radiation in the UV and optical range. Figure 1 shows
a radio map of W 3 superimposed on a Palomar red print. It was obtained
by Sullivan and Downes (1973) with the Westerbork sysnthesis radio telescope
at λ21 cm with an angular resolution of $\alpha \times \delta = 25''\times 28''$ arc. Not shown in
Fig. 1 is W 3(OH), a very compact HII region of angular diameter $1.7''$ arc

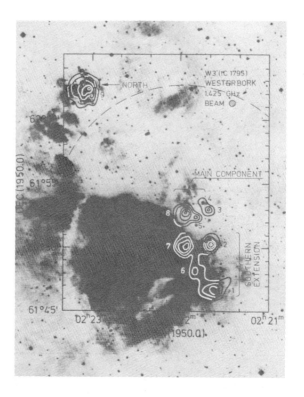

Fig. 1. W 3, main component, southern extension and northern
component observed at λ21 cm with the Westerbork synthesis
telescope. Resolution 25" x 28" arc (α x δ) (Sullivan and Downes,
1973). Nomenclature of source components as used by Mezger and
Wink (1975a).

(Wink et al. 1973), which is located at a distance of about 17' arc to the
south-east of W 3, main component. When observed with higher angular
resolution (some arc sec), both in the IR and at radio wavelengths, W 3
main component (i.e. components 3, 4, 5 and 8 in Fig. 1) breaks up
into a number of individual sources shown in Fig. 2. The IR maps at
λ2.2μm and λ20μm are from Wynn-Williams et al (1972), the aperture
synthesis radio maps at λ3.7 and λ11 cm are from Wink et al (1975). There
are IR sources without radio counterparts and vice versa. In the following,
we refer primarily to W 3(A) which shows a shell-like structure. The
similarity in the brightness distribution maps of W 3(A) at λ2.2μm, and
at radio wavelengths with the map at λ20μm should be noted. The λ2.2μm
and radio radiation originates from free-free and free-bound transitions in
the ionized gas, while the λ20μm radiation is mostly thermal radiation from
dust grains. Figure 3 shows the combined IR and radio spectrum of the two
compact components W 3(A) and W 3(OH). Most of the IR radiation is
emitted at wavelengths > 20μm where no ground-based observations are possible.

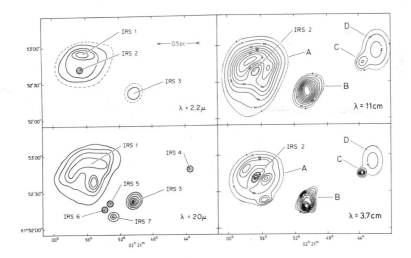

Fig. 2. IR and radio maps of W 3, main component. The IR maps
are from Wynn-Williams et al. (1972) at λ2.2μm and λ20μm. The
radio maps are from Wink et al. (1975) at λ3.7 cm and λ11 cm.

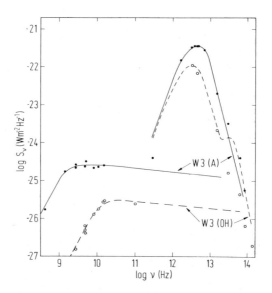

<u>Fig. 3</u>.   Combined radio and   IR spectra of W 3(A)   and   W 3(OH).
Dashed and full lines represent model fits by Krügel and Mezger
(1975).

Recently Fazio et al (1974) resolved W 3(A) and possibly W 3(OH) at
λ69μm. For W 3(A) an angular size of 1.9' arc is quoted, which is
probably an overestimate since contributions from W 3(B, C, D) (see Fig. 2)
are not allowed for. For W 3(OH) an upper limit to the source size < 1' arc,
a peak flux density of 7 x 10³ f.u. and a total flux density of
7 - 15 x 10³ f.u. are given. Comparison with other far IR observations
with lower angular resolution suggests a total flux density at λ69μm of
10 x 10³ f.u. This leads us to the assumption that W 3(OH) is partly
resolved with a 30" arc beam and that its size is in the range 0.5' to
1' arc. However, a direct observational determination of the angular size of
the λ69μm source (which should be possible with an angular resolution of
30" arc) would be of great importance in the context of model fitting (see
Sect. 5).

It is important to realize that, at λ20μm, the sizes of the IR sources are
very similar to the sizes of the corresponding radio sources, i.e. 40" arc
for W 3(A) and ∿ 1.7" arc for W 3(OH). In the far IR, however, the size
of W 3(A) may be somewhat larger (by a factor ∿ 3) at λ69μm than the

corresponding radio source, while W 3(OH) may be larger by a factor 20 to
40 at $\lambda 69 \mu m$ than at radio wavelengths.

3. The Nature of W 3

W 3 appears to be an O-star association in a very early evolutionary stage.
In Fig. 1 we see that its thermal radio radiation is resolved into a number
of compact HII regions. By comparing the sum of the flux densities of these
compact components with the total integrated flux density, we find that, in
the case of the main component, most or all of the radio radiation comes from
the compact components, while in the southern extension, the compact components
account for only $\sim 15\%$ of the total flux density.

It has been found that, in O-star associations, star formation occurs in subgroups which contain typically some 1000 $M_\odot$ of stars. The O-stars appear
to form last, and close to the center of the subgroups. They reach the main
sequence (MS) embedded in a dense shell of gas and dust. At first, all the
stellar radiation is absorbed by dust grains and the shell is observable as
an IR source whose spectrum peaks at about $\lambda 20 \mu m$, and which has no radio
counterpart. At gas densities $n_H < 10^5$ atoms cm^{-3} a compact HII region
can form which is still surrounded by a dense shell of neutral gas. We expect
to observe ionization bounded HII regions with well-defined Strömgren spheres,
which are associated with IR sources whose spectra peak at $\sim \lambda 100 \mu m$. We
suggest that the compact components of W 3 (main component) and W 3(OH)
represent this evolutionary stage.

Once the neutral shell is fully ionized (which requires an expansion of the
HII region), the ionization front progresses quickly into the surrounding
region of lower density. The dust cocoon becomes transparent and part of the
stellar radiation > 912 $\overset{o}{A}$ escapes into the interstellar space. At this
evolutionary stage one expects to find compact HII regions (i.e. condensation
of ionized gas) embedded in an extended low-density HII region. The compact
HII regions, at this stage, would no longer be associated with strong IR
sources. We suggest that the southern extension of W 3 represents this
evolutionary stage.

From IR and radio observations, we can immediately derive some important
integral characteristics of the ionizing star(s) of a compact HII region.
The IR luminosity L_{IR} is a lower limit to the stellar luminosity

$$L_{IR} = 4\pi D^2 \int_{\lambda_1}^{\lambda_2} S_\lambda d\lambda \leq L_*$$

The equality sign holds if all photons get absorbed by dust inside the HII
region or in the surrounding dense shell of neutral gas. The radio flux
density, S_ν, at a frequency where the HII region is transparent, yields
the number N_c' of stellar Lyman continuum photons which are absorbed by
gas

$$D^2 S_\nu \propto N_c' \leq N_c$$

N_c' is a lower limit to the stellar Lyc-photon emission N_c. The equality
sign holds if the HII region is ionization bounded and if no Lyc-photons
are absorbed by dust grains.

While most of the luminosity, especially of the Lyc-photon luminosity, resides
in O-stars, most of the total stellar mass resides in low-mass stars. If
stars of different mass are always formed in the same proportions, it is
possible to derive a mass-to-Lyc-photon ratio. Mezger, Smith and Churchwell
(1974) estimated, for Salpeter's initial mass function, that on the average
2.2E46 Lyc-photons are associated with 1 $M_\odot$ of newly formed stars.
Multiplying the number of Lyc-photons required for the ionization of compact
HII regions in W 3 (main component) by this "mass-to-luminosity" ratio,
one finds that typically some 100 $M_\odot$ to some 1000 $M_\odot$ of stars of various
masses are associated with compact components.

4. Dust Inside the HII Regions

The existence of dust inside HII regions, its effect on the ionization

structure and its contribution to the total IR radiation has become the
issue of a somewhat heated discussion in connection with the interpretation
of the IR radiation from HII regions.

There is no doubt about the existence of dust grains in HII regions (see
e.g. O'Dell and Hubbard 1965, Münch and Persson 1971). Discussion now
centers mainly on the degree of depletion and on the absorption characteristics
of dust grains in the Lyc range.

All Lyc-photons eventually are degraded into Lyman alpha (Lα) photons. It is
assumed that all Lα-photons are absorbed by dust grains inside HII regions.
However, Panagia (priv. communication) has shown that this assumption only
holds if the dust in the HII region is not too greatly depleted. If
Lα-absorption were the only source of heating, the IR-luminosity would be
equal to $N_c' h\nu_\alpha$, with N_c' the number of Lyc-photons absorbed by the gas.
The observable quantity

$$(IR)_{ex} = \frac{L_{IR}}{N_c' h\nu_\alpha} - 1$$

defined as IR-excess radiation, is however always larger than zero (see e.g.
Churchwell et al. 1974). This proves that dust grains also absorb photons
other than Lα-photons, but it does not prove that Lyc-photons are absorbed.

Absorption of Lyc-photons by dust can be inferred in two ways. If the
compact HII regions is surrounded by a dense shell of gas and dust,
$L_{IR} = L_*$. From L_* and stellar model atmospheres, we can estimate the
number N_c of Lyc-photons emitted by the ionizing star. N_c', the number of
Lyc-photons absorbed by the gas is determined from the radio flux density.
If $N_c'/N_c < 1$, as is usually the case, we may conclude that the fraction
$(1 - N_c'/N_c)$ of the intrinsic stellar Lyc-radiation is directly absorbed by
dust grains inside the HII region.

This procedure is not very satisfactory since it involves model atmospheres
and it implies that stellar radiation at all wavelengths predominantly comes
from one single star. The selective absorption of Lyc-photons by dust grains

and its effect on the ionization structure, however, yields results which
allow a direct estimate of the absorption cross sections of dust in the Lyc-
range (Mezger et al. 1974; see also the contribution by L.F. Smith, p. 230
this volume). This will be briefly discussed. Figure 4 shows the
continuum spectrum of a typical O-star of effective temperature 5×10^4K
(which corresponds approximately to an O5 star). This star emits 65%
of its total energy in the Lyc-range. Deviations from a blackbody spectrum
in the Lyc are strong. The emission of photons at wavelengths $< 227 \overset{o}{A}$, the
ionization limit for He^+, is negligible. The amount of He^{++} in "normal"
HII regions (which are ionized by MS stars) is therefore negligible and
the ionization structure of these HII regions can be characterized by the
ratio

$$R = \int_{V(He^+)} n_p^2 \, dV \, / \, \int_{V(H^+)} n_p^2 \, dV \, < \, 1$$

of the volumes of He^+ - to H^+ - region, weighted with the square of the

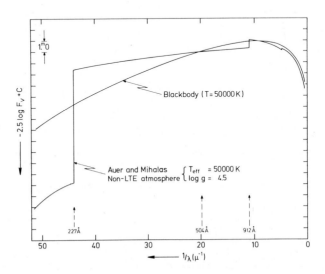

<u>Fig. 4</u>. Comparison of the non-LTE radiation spectrum of an O5-star
(T_{eff} = 50,000 K) with the corresponding blackbody spectrum.

proton density. The abundance y^+ of ionized He, i.e. the ratio of
He^+-ions to H^+-ions, integrated over the H^+-region, can be determined by
observations of adjacent H- and He-recombination lines. y^+ is related
to the total He-abundance y by

$$y^+ = Ry .$$

Between the galactic radii 4-12 kpc, y ≃ 0.10 (Churchwell et al. 1974).
Therefore, observations of y^+ yield immediately the quantity R.

For a given total He-abundance, R depends only on the ratio $\gamma = N_{He}/N_H$ of
He-photons (504 > $\lambda/\overset{o}{A}$ > 227) to H-photons (912 > $\lambda/\overset{o}{A}$ > 504). For y = 0.10,
R = 1 if γ > 0.20; this means R should be unity for stars of spectral
type O9 or earlier (Mezger et al. 1974). Observations show, however, that
for many giant HII regions, which must be ionized by early O-stars, R < 1
(Churchwell et al. 1974; for a recent view see Smith 1975a), e.g. for W 3(A)
we find N'_c/N_c = 0.32 and R = 0.65. Fits to various observations show
that dust grains inside HII regions absorb He-photons 5 to 7 times as
strongly as H-photons, i.e. $a_o = x\sigma_{He}/x\sigma_H \simeq 6 \pm 1$, with x the number of
dust grains per H-atom and σ the absorption cross sections. The absorption
cross section of dust grains for H-photons, however, is remarkably small and
amounts to only 20% of the extinction cross section in the visual. R,
N'_c/N_c and the fraction L'_{IR}/L_* of the total stellar radiation which is
absorbed and re-emitted in the IR by dust inside the HII region can be
expressed as functions of the absorption depth τ_H for H-photons by dust
grains inside the HII region. This is shown in Fig. 5. We see that
for R = 0.65, about 65% of the total IR radiation comes from inside the
HII region, and that about 60% of the intrinsic stellar Lyc-radiation is
directly absorbed by dust grains. This agrees with the observation that
the size of the far IR source associated with W 3(A) is < 3 times the
size of the compact HII region, and that $N'_c/N_c \simeq 0.3$.

5. A Model Fit to W 3(A) and W 3(OH)

One of the puzzling facts is the similarity of most far IR spectra of compact

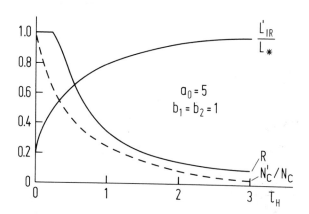

Fig. 5. Diagram showing as a function of the absorption depth τ_H for H-photons inside the HII region: The fraction L'_{IR}/L of the total stellar luminosity radiated in the IR by dust inside the HII region. The ratio R of He-to H-Strömgren spheres, weighted with the square of the proton density. The fraction N'_C/N_C of stellar Lyc-photons absorbed by the gas in the HII region (Mezger and Wink, 1975b).

HII regions (see e.g. Fig. 1). The temperature of the dust grains depends primarily on the distance from the (ionizing) star (see e.g. the temperature profiles T_g in Figures 6 and 7). If the IR radiation of W 3(OH) were from dust grains inside the compact HII region of angular diameter 1.7" arc, its spectrum should peak at $\sim \lambda 20\mu m$. If the IR radiation of W 3(A) were from dust grains uniformly distributed within a sphere of angular diameter $\sim 40"$ arc, the resulting spectrum would be considerably broader since it would be integrated over dust grains with widely differing temperatures.

In the case of these two HII regions, there are sufficient observations available to allow a detailed model fit. Input parameters to this model fit are:

The mass absorption coefficients for He- and H-photons, κ_D^{He} , κ_D^H, and for photons longward of the Lyc-limit, $\kappa_D^{\lambda>912}$; the dust-to-gas mass ratio m_d/m_g; and the absorption efficiency in the IR, $Q = \beta 2\pi a/\lambda$, where $\beta = 8m''/3$ and m'' is the imaginary part of the complex index of refraction of the grain material.

The most important input parameter, however, is the density
profile inside and outside the HII region.

Figure 6 shows the density profile (n_p) which gives the best fit to various
observations of W 3(A). The HII region consists, in essence, of a shell
of ionized gas (n_e = 8000 cm^{-3}) which is embedded in a shell of neutral
gas of equal gas density. The width of the neutral shell must be large
enough to absorb all photons > 912 Å. The ratio of the absorption cross
sections a_o = $x\sigma_{He}/x\sigma_H$ = κ_D^{He}/κ_D^H must be ~ 5 to account for the observed
ratio R = 0.65 of He- to H-Strömgren spheres. However, the fit requires
an increase of the dust-to-gas ratio in the ionized shell by a factor ~ 6
as compared to the average interstellar matter. The shell-like structure
also explains the similar sizes of W 3(A) in the radio range, at $\lambda 20\mu m$ and

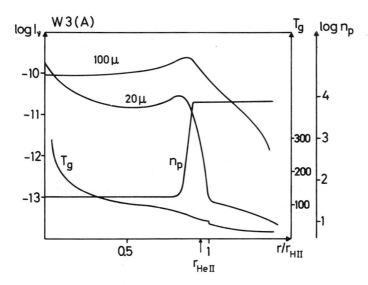

Fig. 6. Cross sections through W 3(A) obtained from model fit:
n_p = proton density (for r/r_{HII} > 1, the curve n_p refers to
the density n_H of the neutral gas); T_g = grain temperature.
$\lambda 20\mu m$ and $\lambda 100\mu m$ refer to the surface brightness at these
wavelengths (Krügel and Mezger, 1975).

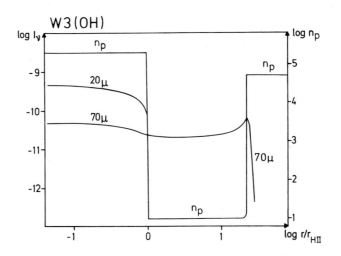

Fig. 7. Cross sections through W 3(OH) obtained from model fit
(Krügel and Mezger, 1975). Notations as in Fig. 6.

at   λ100μm, as shown by the corresponding brightness profiles in Fig. 6. A
shell-structure is also indicated by the observed brightness distributions of
W 3(A) at IR- and radio wavelengths (Fig. 2).

Figure 7 shows the corresponding profiles for W 3(OH). Here the ionized gas
of density 10^5 cm^{-3} is concentrated in a small sphere. Outside this sphere,
the gas density falls off to low values and attains high densities only at
distances of 22 times the radius of the compact HII region. (This distance
is not a very critical model parameter). Important is the fact that the
compact HII region must be very much depleted of dust (by factors ∿ 10^{-2}
compared to the average interstellar matter). The dust-to-gas ratio in the
neutral dense gas shell may be normal or somewhat increased, however, the
model is rather insensitive to a particular choice of this parameter. Due to
the dust-depletion in the compact HII region, only Lα-photons but no
Lyc-photons are absorbed inside the HII region. To explain the low
N_c'/N_c ratio (derived in the manner described above) we have to assume that
the ionizing star of W 3(OH) has not yet reached the MS and therefore
emits less Lyc-photons than would be expected for a MS-star with the same
luminosity. The sizes of the radio and λ20μm sources are similar while

the size of the λ70μm source is about 24 times larger as shown by the brightness profiles in Fig. 7.

6. Implications for the Evolution of Compact HII Regions and for their IR Spectra

W 3(OH) appears to be ionized by an O-star which has not yet reached the MS and therefore has a Lyc-photon flux which is lower than that of a ZAMS star of the same luminosity. This star is embedded in a dust-depleted compact HII region which is surrounded by a dense shell of neutral gas and dust, whose inner radius is twenty to forty times that of the compact HII region. W 3(A) on the other hand, is surrounded by a dense shell of ionized gas and enriched dust, whose diameter is similar to that of the neutral shell surrounding the compact HII region W 3(OH). The ionized shell of W 3(A) appears to be embedded in a dense cloud of neutral gas and ionized by a ZAMS star.

This picture ties in nicely with the dynamical evolution of a compact HII region where dust plays a dominant role. This evolutionary sequence was first outlined by Davidson and Harwit (1967) and subsequently worked out in more detail by Krügel (1974).

When a massive proto-star approaches the main sequence, its surface temperature is low and it cannot ionize the ambient gas, but it has already acquired its full luminosity. The radiation pressure acts on the dust and drives it at high speed ($>>$ 10 km s^{-1}) outwards, without dragging along the neutral gas, because friction between the gas and dust by elastic collisions is weak. The dust piles up in a front until its density there is so high that coupling between the gas and dust becomes effective. The dust front is optically thick at visible wavelengths and hides the star from the optical observer. The dust front takes up the whole momentum of the stellar radiation. Its drift velocity through the gas is small (about 1 km s^{-1}) compared to the speed at which it is driven away from the star (about 10 km s^{-1}). Therefore, the dust front acts like a supersonic piston on the surrounding gas leaving in its wake a highly rarified region.

When the star becomes hot enough, it ionizes (part of) its surrounding gas

from which the dust has been expelled, thus forming a dust-depleted HII
region, which is surrounded by a low-density region (which has been cleared
by the dust front), and further out by the dust front itself. This
may be the stage at which we observe W 3(OH).

In the further evolution, the ionization front overruns the compact HII
region surrounding the star and proceeds rapidly through the rarified gas
until it reaches the dust front. The grains there are exposed to energetic
Lyman continuum radiation and are (probably) charged by the photo-electric
effect. The ambient gas is ionized, and the grains are effectively frozen
into the gas by Coulomb forces. When the compressed material in the dust
front is ionized, its pressure rises and it expands. Now the situation is very
similar to W 3(A): a low density inner region (the central condensation may
have dispersed) surrounded by a dense shell with enhanced dust-to-gas ratio.

At this stage the shell-like HII region is still surrounded by dense
neutral gas and therefore, has a well defined outer boundary. In its further
evolution, more and more of this neutral shell gets ionized while the density
of the HII region decreases. Eventually all of the neutral gas is ionized
and at that stage the HII region appears as a condensation of medium electron
density embedded in an extended low-density HII region. Since the optical
absorption depths in the ionized gas are small, HII regions at this
evolutionary stage are no longer observable as strong IR sources. The
compact HII regions in W 3, southern extension, appear to represent this
evolutionary stage.

If the formation of a dust front is a typical stage in the pre-MS evolution
of O-stars, it provides a simple explanation of the similarity of most far
IR spectra of galactic HII regions; providing the "thermostat which keeps
the grain temperatures at about 70-80 K". The dust surrounding O-stars
would always form a shell of typical radius of the order 10^{18} cm; this radius
may depend somewhat on the stellar luminosity. This is of the right order to
yield the observed colour temperatures of dust grains in and around HII
regions.

References

Churchwell, E., Mezger, P.G., Huchtmeier, W., 1974, Astron.Astrophys. 32,
    283.
Davidson, K., Harwit, M., 1967, Astrophys.J. 148, 443.
Fazio, G.D., Kleinmann, D.E., Noyes, R.W., Wright, E.L., Zeilik II, M.,
    Low, F.J., 1974, Proc. ESRO Symp. HII Regions and the Galactic Center,
    A.M.F. Moorwood, ed.
Krügel, E., 1974, unpubl. Ph.D. Thesis, Univ. of Göttingen.
Krügel, E., Mezger, P.G., 1975, Astron.Astrophys., (in press).
Mezger, P.G., Smith, L.F., Churchwell, E., 1974, Astron.Astrophys. 32, 269
    (Paper I).
Mezger, P.G., Wink, J.E., 1975a, Mem. della Societa Astron. Italiana (in press)
    (Paper II).
Mezger, P.G., Wink, J.E., 1975b, Proc. EPS Symposium HII Regions and Related
    Topics, T.L. Wilson and D. Downes, ed., (in press).
Münch, G., Persson, S.E., 1971, Astrophys.J. 165, 241.
O'Dell, C.R., Hubbard, W.B., 1965, Astrophys.J. 142, 591.
Smith, L.F., 1975, Proc. EPS Symposium on HII Regions and Related Topics,
    T.L. Wilson and D. Downes, ed., (in press).
Sullivan III, W.T., Downes, D., 1973, Astron.Astrophys. 29, 369.
Wink, J.E., Altenhoff, W.J., Webster, W.J., 1973, Astron.Astrophys. 22, 251.
Wink, J.E., Altenhoff, W.J., Webster, W.J., 1975, Astron.Astrophys. 38, 109.
Wynn-Williams, C.G., Becklin, E.E., Neugebauer, G., 1972, Mon.Not.R.Astr.Soc.
    160, 1; 1974, Astrophys.J. 187, 473.

DISCUSSION

Wynn-Williams The low-frequency radio spectrum of W 3(OH) falls off very
steeply, suggestive of a strong cut-off in the density distribution. Your
model of a density-bounded W 3(OH) would surely have a gradually falling
off density distribution as it expands into the vacuum.
Mezger That is correct. We did not do any evolutionary computations for
the HII region but only a model fit to the infrared and radio spectrum.
van Bueren Why does the dust shell become opaque at a certain moment in the
expansion?
Whitworth If you take a given amount of dust further away from the star it
will present a lower optical depth – the optical depth increases as the dust-
shell expands because it sweeps additional infalling dust like a snow-plough.
Mezger For us the most important effect is that with increasing dust-to-gas
ratio the neutral gas gets coupled to the grains. The expanding dust front
thus acts like a piston on the neutral gas.
Bussoletti Is it possible using your model to say something about the amount
of dust present outside the HII region in the cold medium?
Mezger Not much. The models are quite insensitive to the dust density in the
neutral gas.
Fazio What is the angular diameter of the dust shell in W 3(A)? Could it
be as large as 1' arc?
Mezger Yes. The model is quite insensitive to the density of the neutral
gas outside the ionized gas. By lowering the density one gets a larger
diameter for the far infrared source without affecting its spectrum strongly.
Silk An important point in Dr. Mezger's argument is that the far infrared
sources associated with HII regions such as W 3(OH) are considerably
more extended than the regions of peak radio continuum emission. Obviously,

if one had a distribution of B stars that produced relatively little
ionization, one could obtain an additional and more extensive energy input
than is produced internally in the HII region. This would affect any
conclusions about dust in the HII regions. I would therefore like to ask
the infrared observers: is there any general and convincing evidence that
supports the picture that the infrared emission is produced in a more extended
region? Both multicolour and high resolution observations are urgently needed.
so as to detect colder dust away from the HII region.
<u>several voices</u> No!

A COMMENT ON THE SIMILARITY IN THE SPECTRA
OF FAR INFRARED SOURCES

H. Okuda

It is interesting to note the remarkable similarity in the spectra of many far infrared sources all of which peak near 100 μm. The emission is considered to be thermal radiation of grains heated by some sort of primary source.

The spectral shape of the thermal emission in general depends strongly on the temperature of the grains. Since the same peak wavelength has been observed in variety of sources[1] - HII regions, the Galactic centre and some extragalactic nuclei - they are not likely to be due to emission by grains with identical temperatures. The observed behaviour could be attributed more reasonably to a material dependent property especially since the source is optically thin, as is indeed the case for most sources in far infrared regions.

As an example, the thermal spectra of ice grains have been calculated using the data given by Irvine and Pollack[2] for the temperatures of 30, 60 and 100 K. The results are shown in Fig.1. In spite of the relatively wide temperature range considered here, the spectral peaks show much less deviation, almost concentrated in 50 ~ 70 μm. Since ice grains can survive in a vacuum environment only at temperatures lower than 100 K and a realistic spectrum involves a spread of temperatures, a mixture of these curves would more closely resemble the observed spectral shape.

Although the ice grains are not considered to be the dominant constituent of the interstellar grains[3], the ice or icy grains may play an important role in the far infrared. More information on the optical properties of various candidate materials is urgent for further discussion in the far infrared region.

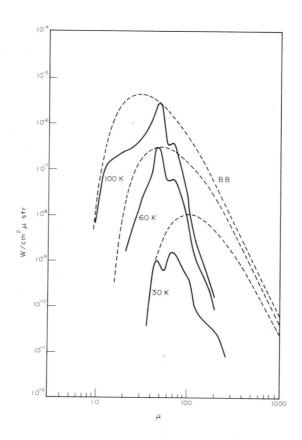

<u>Fig. 1.</u>   Spectrum of thermal emission of ice grains (size = 1 μm)
at temperatures of 30, 60 and 100 K.   Black body spectra for
corresponding temperatures are shown by dotted lines for
comparison.

Acknowledgment

This work was completed during the author's stay in the Department of Applied
Mathematics and Astronomy, University College, Cardiff.

References

(1)   Harper, D.A., and Low, F.J., 1973, <u>Astrophys.J.</u> 182, 489.
(2)   Irvine, W.M. & Pollack, J.B., 1968, <u>Icarus</u> 8, 324.
(3)   Knacke, F.F., Cudaback, D.D. & Gaustad, J.E., 1969, <u>Astrophys.J.</u> 158, 151.

PART 6

THEORETICAL MODELS

OPTICAL DEPTHS OF FAR INFRARED SOURCES

C. D. Andriesse

Kapteyn Astronomical Institute, Groningen, The Netherlands

Abstract The optical depths at 350 µm of far-infrared sources
are discussed in the context of collapsing clouds. It is assumed
that the opacity is determined by dust particles, for which the
far-infrared properties can be reliably estimated. The sources
are found to be optically thin at 350 µm ($\tau \overset{\sim}{\sim} 0.001$) and reasonable
values for their mass are derived. The complete infrared spectrum
of Sgr B2 is analysed, giving $\tau_{350} < 0.03$ and $M \overset{\sim}{\sim} 10^7 M_\odot$.

The point has been made that the far-infrared sources in our Galaxy could be
optically thick. If this were true, the recent 350 µm data would lose much
of their significance for mass determinations. We discuss the far-infrared
optical depth in its astrophysical context and scrutinize the case of Sgr B2.

1. Consider a spherical source with radius R containing a mass M of gas
and dust. The mass fraction β ($\overset{\sim}{\sim} 10^{-2}$) of the dust can be taken constant
and the mass absorption coefficient κ_m in the far-infrared is essentially
given by the dust. The optical depth in this source is at most

$$\tau = \kappa_m \frac{3 \beta M}{4 \pi R^2} \tag{1}$$

The likely far-infrared properties of dust particles are investigated by the
author (1974), who finds

$$\kappa_m = \frac{A \omega^2}{\rho} \tag{2}$$

with ω the angular frequency and ρ ($\overset{\sim}{\sim} 10^3$ kg m^{-3}) the specific density of a solid. The strength A depends on the wavelengths λ_n of the lattice resonances and on the ratio ε = (thermal energy)/(binding energy), viz. $A = 7/(3\pi^2 c^2)\, \varepsilon^{\frac{1}{2}}\, 1/n \, \Sigma \, \lambda_n$. Loose bonds will make ε of the order of 10^{-2}. whereas $1/n \, \Sigma \, \lambda_n$ may attain 30 μm. The order of magnitude of A is then $10^{-23}\, m^{-1}\, s^2$, with which it is possible to derive a realistic temperature for interstellar dust.

To discuss M one can adopt the view that most - if not all - far-infrared sources originate from gravitational collapse of a cloud. In spiral arms M will be a little below 10^{34} kg (the canonical value is 3000 $M_\odot$). For a typical radius of 1 pc one finds then $M/R^2 = 10$ kg m^{-2}, but the question is whether 1 pc is typical for the extent of the far-infrared sources. Consider the condition for collapse, stating that the inward gravitational plus thermal pressure exceeds the outward magnetic plus thermal pressure. When the thermal pressures cancel each other (case of the quasi-equilibrium in the two-phase model of the interstellar medium; no shock), the condition reads, in terms of energy

$$\frac{4}{3} \pi R^3 \frac{B^2}{2\mu_o} < G \frac{M^2}{R} \quad . \tag{3}$$

Here B is the magnetic induction ($\tilde{} \; 3 \times 10^{-10}$ weber m^{-2} in spiral arms), μ_o the permeability of vacuum and G the gravitational constant. As soon as (3) is satisfied, e.g. after the encounter and merging of subcritical clouds, collapse will ensure. Then both terms will be nearly equal, or

$$\frac{M}{R^2} = B \sqrt{\frac{2\pi}{3\mu_o G}} \quad . \tag{4}$$

For the above magnetic induction this gives 5×10^{-2} kg m^{-2}. The values of B and R apply both to the initial cloud; when the magnetic flux is conserved during the collapse, B grows with $1/R^2$. For the guessed ratio M/R^2 of 10 kg m^{-2}, the induction would be 10^{-7} weber m^{-2} (as in the Crab Nebula), which is unlikely. Therefore R will be some 10 pc, the size of a "standard" cloud still. We recall that the outer regions of a collapsing cloud have not yet moved far inward at the moment, that the runaway increase of the central density has led to the energy production, responsible for the far-infrared radiation (Larson, 1973).

Allowing for a decrease of R by a small factor f ($\stackrel{<}{\sim}$ 5), one has

$$\tau = \frac{3 \beta f^2}{4\pi} \sqrt{\frac{2\pi}{3\mu_o G}} \frac{AB\omega^2}{\rho} \quad .$$ (5)

At the wavelength of 350 μm ($\omega = 5.4 \times 10^{12}$ s^{-1}) we get from the above discussed values the estimate $\tau_{350} = 10^{-3} \ll 1$.

2. The above result enables us to derive the mass M from measured fluxes at 350 μm in a very simple way. The emission efficiency of a dust particle with the average radius a is obtained from (2) by

$$Q = \frac{a}{3} \rho \kappa_m \quad ,$$ (6)

which differs from the dilution factor discussed by Andriesse & Olthof (1973). Assume that all dust particles in the source have the same temperature T (this can be founded on an emission law $\sim T^{6.5}$, which makes T rather independent on variations in the radiation field that heats the dust particles). N particles in an optically thin source at distance r give the flux

$$F_\omega = \frac{\beta M \kappa_m}{4\pi r^2} B_\omega(T) \quad ,$$ (7)

where βM has been substituted for $N \frac{4}{3} \pi a^3 \rho$ and $B_\omega(t)$ is the Planck function. When T > 30 K one obtains from (7) and (2) in the Jeans approximation

$$M = \alpha r^2 F_\omega \quad ,$$ (8)

$$\alpha = \frac{8\pi^3 c^2 \rho}{A\beta kT\omega^4} \tag{9}$$

where c is the velocity of light and k Boltzmann's constant. Because T enters linearly and only orders of magnitude count, we put T = 50 K and obtain for 350 μm the value $\alpha = 4 \times 10^{17}$ kg joule^{-1}.

Consider the far infrared sources of Orion and Sgr B2. For the Orion source at $r = 1.5 \times 10^{19}$m, with a 350 μm-flux $F = 5 \times 10^{-23}$ Wm^{-2} Hz^{-1}, one obtains $M = 4.5 \times 10^{33}$kg (or 2300 $M_\odot$). This is much more than derived earlier * , but not far from the canonical value for collapse in spiral arms. For Sgr B2 at $r = 2.5 \times 10^{20}$m, with a total 350 μm-flux F close to 10^{-21} Wm^{-2} Hz^{-1}, one obtains $M = 2.4 \times 10^{37}$kg (or $10^7 M_\odot$). Molecular line data for this cource imply a mass of this order also ($> 3 \times 10^6 M_\odot$, Zuckerman & Palmer, 1974). We can conclude that the derived masses are as large as they should be, which is an indirect proof of a small optical depth.

3. For Sgr B2 we verify whether $\tau_{350} \ll 1$ is compatible with all existing infrared data. It is for this source that the conjecture was made that $\tau_{350} \approx 1$. In Fig. 1 a plot is given of the radio continuum (R), the 3.5 mm flux (HMM) measured by Hobbs et al. (1971), the 0.8-3.0 mm flux (CAR) measured by Clegg et al. (1974), the 300-400 μm flux (RSJG) measured by Righini et al. (1975), the 84-130 μm flux (O) measured by Olthof (1974) and the 17-24 μm flux (LL) measured by Lemke & Low (1972). Whereas at (O), (LL) and probably (HMM) the total flux from the source is measured, one has to correct for beam effects at (CAR) and (RSJG). This correction is possible by virtue of the source profile at 350 μm measured by Righini et al. The flux at (RSJG) in a diaphragm of 5' arc (small cross) then has to be increased by a factor 2 and similarly the flux at (CAR) in 1' arc by a factor 10; in the latter case the uncertainty is increased by the rather small

---

* At first sight there seems to be a conflict with the mass ratio to ionized gas (>> 1); however, when ultraviolet photons cannot penetrate deeply in the source, the mass in ionized hydrogen that can be derived from radio observations will only be a fraction of the total mass; Pottasch (1974) argued that the extent of sources in the radio- and near-infrared region is ultraviolet-photon bounded rather than material bounded; the larger extent of these sources at 100 μm proves that far reaching infrared photons are responsible for heating of the dust outside the near-infrared node(s).

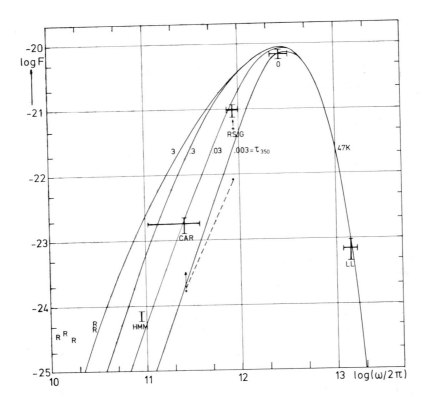

<u>Fig. 1</u>. The infrared and radio spectrum of Sgr B2 (crosses)
compared with theoretical models with different optical depths at
350 μm.

beam-switch of 2' arc that was used.

The full curves shown in Fig. 1 give the theoretical

$$F_\omega = \left(\frac{R}{r}\right)^2 B_\omega(T)[1 - e^{-\tau}] \tag{10}$$

for τ given by (1) and (2). The temperature of 47 K is chosen between
the colour temperature 44 K and the brightness temperature 51 K given by
Lemke & Low for 21 μm, defining in that way an effective R of 9×10^{16} m.
The behaviour of F_ω for $\omega/2\pi > 5 \times 10^{12}$ s^{-1} is that of a blackbody. For

smaller frequencies we obtain a satisfactory fit for $\tau_{350} = 0.03$. However, the source is larger here than 9×10^{16} m (1'.3 arc), so that the actual depth may be smaller still by an order of magnitude. There is an obvious difficulty in describing F_ω by (10). One either has to assign a larger extent to the source than observed at 21 μm (and thus assume a lower temperature, close to 30 K) or a smaller extent than observed at 350 μm. This artificial procedure is due to the assumption of a constant temperature.

Finally, a direct comparison of the central 1' arc data at (CAR) and (RSJG) (indicated by the small circle) would suggest an ω^3-law (dashed line in Fig. 1), which seems to contradict $\tau_{350} \ll 1$. In view of the observational effects of small beams in large sources, a conclusion that $\tau_{350} \overset{\sim}{\sim} 1$ is not fully warranted. On the contrary, putting all pieces of evidence together, we think that it is reasonably certain that $\tau_{350} \ll 1$.

Acknowledgements

It is a pleasure to acknowledge the help of G. A. van Moorsel in preparing Fig. 1.

References

Andriesse, C.D., 1974, Astr.Astroph., 37, 257-262.
Andriesse, C.D., Olthof, H., 1973, Astr.Astrophys., 27, 319-321.
Clegg, P.E., Ade, P.A.R., Rowan-Robinson, M., 1974, Nature, 249, 530-532.
Hobbs, R.W., Modali, S.B., Maran, S.P., 1971, Astrophys.J., 165, L87-L93.
Larson, R.B., 1973, Ann.Rev.Astr.Astroph., 11, 219-238.
Lemke, D., Low, F.J., 1972, Astrophys.J., 177, L53-L57.
Olthof, H., 1974, ESRO SP-105 (HII regions & Galactic centre), 235-238.
Pottasch, S.R., 1974, Astr.Astrophys., 30, 371-379.
Righini, G., Simon, M., Joyce, R.R., Gezari, D.Y., 1975, Astrophys.J., 195, L77-L79.
Zuckerman, B., Palmer, P., 1974, Ann.Rev.Astr.Astroph., 12, 279-313.

DISCUSSION

Gilra (i) You did not emphasise that the expression you wrote for the absorption cross-section is for non-metallic particles. (ii) In your formula you probably do not consider small particles. In that case one has the shape effects involved. I have derived analytical expressions and in the extreme cases the difference is a factor of ε_o^2, where ε_o is the static dielectric constant, which could be a factor of 10 to 100.
Andriesse I only considered non-metallic grains, and for good reasons. The proof of the pudding is in the eating: with metallic grains you would not get a reasonable fit to the observed spectral distribution of the infrared

emission, nor a reasonable value for the mass. Concerning your remark
on a possible high static dielectric constant in small elongated particles,
I think that there are no compelling reasons to adopt such extreme shapes,
to account either for the polarization data or for the far infrared emission.
Furthermore one should not invoke whisker-type shapes in too much of a hurry,
because they are difficult to grow from a complex mixture of gases in cosmic
abundances. I want to refer here to a discussion in the appendix of my
recent paper on grains and to the recent work by Dr. B. Donn (20th Liege
Astrophysics Colloquium, 1975) on cosmic sublimation and grain growth. But
this is already a comment on the forthcoming contribution of Dr. Edmunds.
Greenberg If shape effects (elongation, flattening) on absorption by grains
is to increase efficiency by as much as a factor of 5, I do not expect
Dr. Andriesse's conclusion to be changed. But if metallic rather than
dielectric particles are assumed there could perhaps be a significant effect.
As a matter of fact, for dielectric particles the shape factor is probably
$\lesssim 2$.

FAR INFRARED RADIATION FROM DUST WITHIN HII REGIONS

Per A. Aannestad

University of Arizona, Tucson, Arizona 85721, U.S.A.

Abstract A spherical model of dusty HII regions containing core-mantle grains where the cores are evaporated in the innermost region and the mantles are evaporated in an outer region has been compared with the observations. It is found that effective dust optical depths in the Lyman continuum in the range 0.3 to 3 with a mean value about unity can explain the observed correlation between the total infrared luminosity and the radio continuum flux density. Employing the optical properties of olivine core and olivine core-ice mantle grains and assuming evaporation temperatures of 1000 K and 100 K respectively, infrared spectra have been computed for the HII regions of Orion A, W3, and M17. Comparison with observations shows reasonable agreement when account is taken of the differing beam widths and chopping offsets in the observations.

1. Introduction

It is now well established that many of the far infrared sources are sources of heated dust connected with galactic HII regions. However, it is not yet clear whether the dust responsible for most of the radiation is situated in neutral gas outside the HII region or exists within the ionized region itself. Recent observations (Emerson et al. 1973, Harper 1974) show that in some cases the far infrared maps correlate closedly with the radio continuum maps, indicating that the dust in these cases is mixed with the ionized gas within the nebulae. In the Trapezium region of the Orion Nebula, observations at 10 to 20 μm (Ney and Allen 1969, Ney et al. 1973) show that silicate dust exists close to the ionizing stars, and observations of the scattered visual continuum radiation (O'Dell et al. 1966, Munch and Persson 1971) show that the dust is well mixed with the gas over a large part of the nebula. It is also indicated that the dust-to-gas ratio is much smaller in the inner regions (at ∿0.1 pc) than in the outer regions (at ∿1 pc) and there may be a "hole" in the dust distribution at the center of the nebula (Simpson 1973).

Dust within HII regions may be original interstellar particles that received new or additional growth in the dense phase prior to the formation of the HII region and its exciting stars. One component of such dust may be silicate cores with an ice or modified ice mantle (Greenberg and Hong 1974, Aannestad 1975). As soon as the ionizing stars turned on, the dust would have been subject to destructive processes such as evaporation and sputtering. The refractory component of the dust can withstand the harsh environment much better than the mantle material and will be subject mainly to evaporation in the immediate neighbourhood of the exciting stars. Farther from the central stars, the temperature of the dust may be low enough for the grains to keep their original mantle. Relative to the lifetime of ordinary HII regions, ice grains slightly hotter than 100 K will rapidly evaporate, whereas slightly cooler grains will survive the evaporation process. Ice grains at 100 K evaporate in $\sim 10^5$ years (Isobe 1970).

Destruction due to sputtering depends on the charge of the grains. The grains may be positively charged by photoejection and thus effectively be shielded from ion sputtering (Mathews 1969). With a sputtering yield for neutral H and He atoms of 0.02 per incident H-atom (Aannestad 1973), the rate of destruction of the ice mantle is $\sim 5 \times 10^{-8}$ $n_e x_e$ μ/year, where n_e is the electron density and x_e is the fraction of neutral H atoms in the HII region. For $x_e = 10^{-3}$ and $n_e \leq 10^3$ cm^{-3}, the time scale for destruction of an ice mantle from 0.15 μm radius to 0.05 μm radius is $\geq 2 \times 10^6$ years, which is longer than the lifetime of most HII regions. However, sputtering due to photon impact may be more important than physical sputtering. The rate of destruction due to this process is $\sim 1.5 \times 10^{-12}$ YF μm/year, where Y is the yield (molecules/photon) and F is the uv photon flux. Y is not well known, but laboratory measurements indicate that it is 10^{-6} to 10^{-7} (Greenberg 1973). Photosputtering is therefore less important than evaporation for the destruction of an ice mantle as long as $F < 10^{12}$ to 10^{13} uv photons cm^{-2}s^{-1}, which corresponds to most of the nebular volume in many cases.

In the following we assume a simple but idealized model for dusty HII regions suggested by the data and the considerations above. The model is similar to

recent models of dusty HII regions discussed by Wright (1973), Panagia
(1974, Pottasch (1974) and de Jong et al. (1975). However, here we consider
two dust components, the core grains and the core-mantle grains, with their
relative distribution determined by the evaporation temperatures, i.e. by the
grain materials in a self-consistent manner.   The grains are assumed to be
silicate cores with ice mantles, and we employ recent calculations of the
optical properties of olivine and olivine core-ice mantle grains (Aannestad
1975).   We also apply this model to the HII regions Orion A, W3 and M17,
and compare the results with observations. The details of the model
calculations will appear elsewhere.

2.   Theoretical Model

The idealized structure of the HII region is shown in Fig. 1.   It is
assumed to be spherical with a radius R and to have a uniform electron
density given by

$$n_e \ (cm^{-3}) \quad = \quad \frac{2.8 \times 10^4}{\theta^{3/2}} \quad (S/D)^{\frac{1}{2}} \tag{1}$$

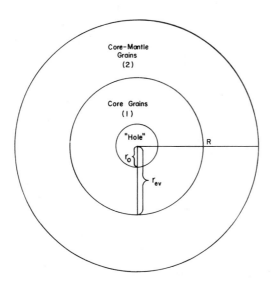

Fig. 1.   Schematic presentation of the dust distribution in the
model HII region.

where S is the observed 5-GHz flux density in flux units, D is the
distance in pc, and θ is the angular diameter in minutes of arc (Schraml
and Mezger 1969).   The electron temperature is 7000 K. Except for the
central "hole", the dust responsible for the infrared emission is
coincident with the HII region. All the Lyman continuum radiation is
assumed to be absorbed either by the gas or by the dust (radiation bounded),
and the dust-to-gas number density $x_g = n_D/n_e$ is assumed to be a constant.
The free parameters are the total effective dust optical depth in the Lyman
continuum τ_D and the ratio of the absorption cross sections in the Lyman
continuum for the core-mantle grains to that of the core grains
$z = \sigma_{D,2} / \sigma_{D,1}$.   In the calculations we assume the absorption efficiency
to be unity throughout the ultraviolet and to be the same for core and core-
mantle grains.   The inner boundaries of the two dust regions r_o and r_{ev}
are determined by the evaporation temperatures of the dust materials.   We
assume that the olivine cores evaporate at 1000 K and that the ice mantles
evaporate at 100 K.

The sources of energy that have been included as input to the nebular grains
are (a) Lyman continuum photons from the central star(s), (b) stellar photons
longward of the Lyman limit, (c) diffuse Ly-α photons, and (d) Ly-α photons
escaping from the inner 'hole'. Using the on-the-spot approximation and an
effective absorption optical depth (Petrosian and Dana 1975), the equation of
transfer for the flux of (a) has been solved analytically for our model HII
region. The photons (b) and (d) are mostly affected only by the dust in the
nebula, and the fluxes therefore have the usual simple exponential variation.
The average flux of the diffuse Ly-α photons is approximately $1/4\alpha\, n_e^2 L$,
where α is the hydrogen recombination coefficient to upper levels, n_e is the
electron density, and L is the mean free path for destruction (escape or
absorption). We have approximated L by $R(1/f + R\, n_d\, \sigma_d)^{-1}$, where f = 2x
ln $\tau_{g,0}$ and $\tau_{g,0}$ is the gas optical depth at the center of the Ly-α line.

If we consider optical depths τ_D of the order of unity, the dust is
optically thin to its own radiation, and the dust temperature T_g is given by
the integral equation

$$4\pi a^2 \int_0^\infty \pi B_\lambda (T_g) Q_\lambda d_\lambda \;=\; \pi a^2 [Q_L F_L(\tau) + Q_B F_B(\tau) + Q_\alpha F_H(\tau)]$$

$$+ \; 4\pi a^2 Q_\alpha F_\alpha, \tag{2}$$

where τ is the independent variable $0 \leq \tau \leq \tau_D$: a is the grain radius: F_L, F_B, F_α, and F_H denote the fluxes of the photons (a), (b), (c) and (d) respectively; and Q denotes the grain absorption efficiencies for the corresponding photons.   The boundaries of the two dust regions are found from Eq.(2) by a simultaneous solution for the two cases $\tau = 0$ (corresponding to $r = r_0$), $T_g = 1000$ K, $a = a_{core}$, $Q_\lambda = Q_\lambda$ (core grains) and $\tau = \tau_{ev}$ (corresponding to $r = r_{ev}$) $T_g = 100$ K, $a = a_{mantle}$, $Q_\lambda = Q_\lambda$ (core-mantle grains).   Once the temperature distribution is obtained, the intrinsic infrared spectrum is found by integrating over the volume of the two dust shells, and the total infrared luminosity is given by an integration over the spectrum.

3.   Infrared Fluxes and Radio Fluxes

It was demonstrated by Harper and Low (1971) and by Emerson et al. (1973) that there is an approximate linear relationship between the broad-band infrared flux and the radio continuum flux density, and that the total infrared energy is approximately five times larger than the energy available from Ly-α photons alone.   The larger sample of objects observed by Furniss et al. (1974) shows a larger scatter, but there still is a general correlation.   It is of interest to estimate the values of the free parameters τ_D and z in our model that will account for the observed points in particular since this procedure is independent of knowing the infrared emissivities of the grain material.

The contribution to the infrared flux by Lyman-continuum photons is given by

$$F_{IR,L} \; (Wm^{-2}) \; = \; 10^{-40} \; \frac{<h\nu>_L f_D N_0}{4\pi D^2} \; , \tag{3}$$

where f_D is the fraction of the total number of Lyman continuum photons N_0 that is absorbed by the dust $<h\nu>_L$ is the average Lyman continuum photon energy in ergs, and D is the distance in pc.   If we assume $<h\nu>_L = 2<h\nu>_{Ly-\alpha}$, which is appropriate for most O stars (Harper 1974) one may show that Eq.(3) becomes

$$F_{IR,L}(W \; m^{-2}) \; = \; 3 \times 10^{-11} (1/f_g - 1)S, \tag{4}$$

where f_g is the fraction of continuum photons absorbed by the gas and S is the 5-GHz flux density of the HII region. In our model f_g is given by

$$
1/f_g = \frac{3z^3}{[\tau_D + \tau_0 + (z-1)\tau_{ev}]^3} ([(\tau_{ev} + \tau_0 - 1)^2 + 1]e^{\tau_{ev}}
$$

$$
- 1 - (\tau_0 - 1)^2 + z^{-3}([(\tau_D - \tau_{ev} + \tau_v - 1)^2 + 1]e^{\tau_D}
$$

$$
- [(\tau_v - 1)^2 + 1]e^{\tau_{ev}} + \frac{1}{3}\tau_0^3) .
\tag{5}
$$

Here $\tau_0 = n_D \sigma_{D,1} r_0$, $\tau_{ev} = n_D \sigma_{D,1} r_{ev} - \tau_0$, and $\tau_v = z(\tau_0 + \tau_{ev})$. For the photons longward of the Lyman limit we similarly obtain

$$
F_{IR,B} \ (W \ m^{-2}) = 3 \times 10^{-11} B \frac{\langle h\nu\rangle_B}{\langle h\nu\rangle_L} (\frac{1 - e^{-\tau_D}}{f_g}) S,
\tag{6}
$$

where B is the ratio of the total number of photons longward of the Lyman limit to N_0, and $\langle h\nu\rangle_B$ is the average energy of such photons. In the calculations the quantity $B\langle h\nu\rangle_B/\langle h\nu\rangle_L$ has been set equal to unity, appropriate for spectral classes 05 through 07, but it may increase rapidly for later spectral classes (Harper 1974).

The contribution from the diffuse Ly-α photons to the infrared flux can be written as

$$
F_{IR,\alpha} = 1.42 \times 10^{-11} S [\frac{\tau_0 + \tau_{ev} + z^{-1}(\tau_D - \tau_{ev})}{1/f + \tau_0 + \tau_{ev} + z^{-1}(\tau_D - \tau_{ev})} (x_{ev}^3 - x_0^3)
$$

$$
+ \frac{\tau_D - \tau_{ev} + z(\tau_0 + \tau_{ev})}{1/f + \tau_D - \tau_{ev} + z(\tau_0 + \tau_{ev})} (1 - x_{ev}^3)] ,
\tag{7}
$$

where $x_0 = r_0/R = z\tau_0/[\tau_D + z\tau_0 + (z-1)\tau_{ev}]$ and $x_{ev} = r_{ev}/R = z(\tau_0 + \tau_{ev})/[\tau_D + z\tau_0 + (z-1)\tau_{ev}]$. In the calculations we have set $f = 10$ (O'Dell 1965).

Finally, the contribution due to the Ly-α photons escaping from the
innermost region is

$$F_{IR,H} = 1.42 \times 10^{-11} \, x_0^3 \, (1 - e^{-\tau_D}) S. \tag{8}$$

The sum of the contributions (4), (6), (7) and (8) is plotted in Fig.2 for
a range of values in τ_D. The parameter z is taken to be 3, but the linear
relations shown are not sensitive to this value, the ordinate decreasing by
about 10% if z is allowed to be as high as 10. The values for τ_0/τ_D and
τ_{ev}/τ_D are typical for the silicate core-ice mantle grains, and again the
plotted relations are insensitive to variations in these values. Also plotted
is the predicted linear relation if only diffuse Ly-α photons are included
as the source for heating the dust   (Eq.(7)). Fig.2 also shows observed
values of the infrared flux versus the radio flux density.   Most of the

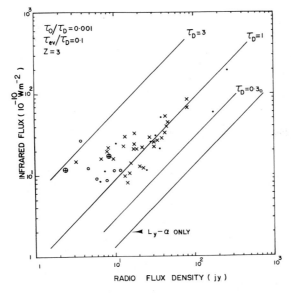

Fig. 2 Predicted linear relations between the total infrared flux
and the radio flux density for various values of the effective
Lyman continuum dust optical depth τ_D. Lower line illustrates
the relationship if only Ly-α photons are assumed to contribute to
the infrared luminosity. Also shown are the observed values for
HII regions separated into four categories as discussed in the text.

infrared fluxes are the peak fluxes in the wavelength interval 40 to 350 μm
from Furniss et al. (1974) but increased by 40% to take account of any energy
outside their wavelength range.   The radio flux densities are from the
tabulation by Furniss et al. but have been reduced to the flux density at a
common frequency of 5 GHz by assuming a spectral index of 0.1.   We have
also included the 45 to 750 μm measurements of DR21 and M17 by Harper and Low
(1971) with the radio data taken from Schraml and Mezger (1969).

The basis for the derived relation between F_{IR} and S is that the infrared
and the radio measurements refer to identical regions of space. Such
observations are often not available, and we have instead attempted to
separate out any size effects by plotting the observations in four different
categories. The filled circles are the sources where the radio size ≤ i.r.
size and may be expected to follow the predicted relation or tend to fall to
the left in the diagram. The observations marked as x are sources where
the observed radio size $\lesssim$ i.r. beam size (4' arc).   In principle, such points
could scatter left or right. The circles are sources where the radio size
> i.r. beam size and such points may be expected to scatter to the right in the
diagram.   Circles with crosses are sources for which a radio size is not
available and could also scatter left or right in the diagram. Excluded from
the plot are NGC 6334, which has no reliable estimate of the total radio flux,
and the Galactic Center region.

We see from Fig. 2 that there is no obvious separation between the different
categories of observed points, indicating that effects other than beam size
differences are more important for the observed scatter. We see that a
variation in τ_D from 0.3 to 3 with a mean value of about unity can adequately
account for the observed scatter. Unfortunately, the effect of a variation in
spectral class of the exciting stars is to mimic the effect of a variation in
τ_D . However, if most of the ionizing stars are assumed to be of earlier
spectral class than about O7, the variation due to spectral class is small
and cannot alone account for the spread in the observations. We conclude that
a model where dust and gas coexist within the HII regions, with an effective
optical depth in the Lyman continuum in the range 0.3 to 3, may account for
the observed correlation between the infrared flux and the radio flux and
also explain the observed scatter.

4. Intensity Profiles and Spectra

We have applied our model to three of the most observed HII regions,
Orion A, W3, and M17. The radio parameters adopted for these regions are
shown in Table 1.

TABLE 1.   RADIO PARAMETERS.*

HII region	D (pc)	R (pc)	θ (arc min)	S (jy)	n_e (cm^{-3})	M_{HII} ($M_\odot$)
Orion A	500	0.6	8.3	400	1000	20
W3	3000	2.4	5.5	68	330	430
M17	1800	2.3	8.6	500	580	630

*D = distance to the HII region;   R = radius; θ = angular diameter;

S = flux density at 5 GHz; n_e = total electron density;

M_{HII} = mass of ionized hydrogen.

Fig. 3 shows the variation of the dust temperature with distance from the
center of Orion A, but the curves are also representative for the other HII
regions.   The optical depth τ_D has been set to unity, and the grain core
size is assumed to be 0.05 μm.   The solid curve shows the temperature
structure when the olivine core-ice mantle grains have a total radius of
0.1 μm, the dashed curve is for a total radius of 0.15 μm, and the dotted
curve is also for a radius of 0.15 μm but with a maximum enhanced far
infrared (40 to 1000 μm) emissivity following a λ^{-1} law for both the core and
the mantle material (Aannestad 1975).   We note that the drop in grain
temperature at the boundary of the core and the core-mantle regions is only
at most 20%, and that variations in mantle sizes or far infrared
emissivities have only a minor influence on the overall temperature
structure. The characteristic features of the curves in Fig. 3 are the very
rapid decrease of the temperature with increasing distance in the innermost
regions and the subsequent slow decrease of the temperature over most of the
nebula.   About 90% of the nebular grains have a temperature between 80 and
50 K.

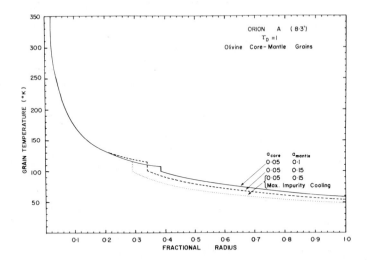

Fig. 3. Temperature variation with distance for olivine core-ice
mantle grains in Orion A (radio diameter 8.3' arc). Olivine cores
are assumed to have a radius of 0.05 μm. Solid curve is mantle
radius of 0.1 μm. Dashed curve is mantle radius of 0.15 μm.
Dotted curve is mantle radius of 0.15 μm but with a maximum
enhancement of the far infrared emissivities for both core and
mantle material. Evaporation temperatures are 1000 K and 100 K
for the olivine cores and the ice mantles, respectively.

A temperature variation as in Fig.3 would be expected to lead to a bright
inner core at short wavelengths (< 20 μm) and a much broader and diffuse
structure at long wavelengths (> 50 μm). This is evident in Fig.4, which
shows the normalized intensity profiles for various wavelengths for the case
of Orion A with τ_D = 1 and grains of core size 0.05 μm and mantle size
0.15 μm. In Fig. 4(a), we note the rapidly changing structure as the wavelength changes from 18.5 μm to 34 μm. Also, the much flatter profile at
34 μm implies that chopping at large offsets (several minutes of arc) may be
necessary to estimate the true flux at such wavelengths. The intensity peak
at about 1.4' arc is due to the onset of the core-mantle region, within which
the intensity then drops off rapidly with increasing angular distance from
the center. In Fig.4(b) the profiles are shown for wavelengths 50 to 300 μm.
We note that the profiles are all similar and peak at the boundary between the
core grains and the core-mantle grains. In order to resolve such a peak, one
would need a resolution of about 30" arc at far infrared wavelengths chopping

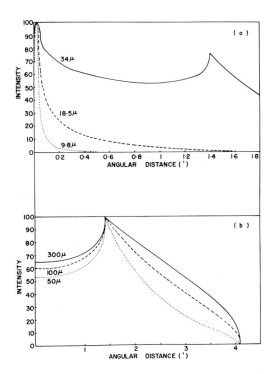

Fig. 4 Normalized intensity profiles as functions of angular
distance from center at various wavelengths for Orion A. Olivine
core-ice mantle grains have a mantle radius of 0.15 μm and a core
radius of 0.05 μm. Dust optical depth τ_D equals 1.

at large offsets. However, any dust density variations and deviations from
spherical symmetry in a real nebula would tend to "wash out" this structure.

Fig. 5 shows calculated spectra for the Orion A region. The dust grains have
a core size of 0.05 μm and mantle size of 0.15 μm. The solid line employs
laboratory measurements of the optical constants for the olivine and the ice
materials up to 250 μm, whereas the dashed curve assumes the maximum
enhancement of far infrared emissivities. Also plotted are observed values
of the infrared flux densities of the Orion A region. In the far infrared
the observed flux is dominated by the contribution from the nearby Kleinmann-
Low Nebula, and in Fig. 5 the vertical dashed lines show a reduction in the
observed values (except for the 90 μm point) by a factor of 3, in an attempt

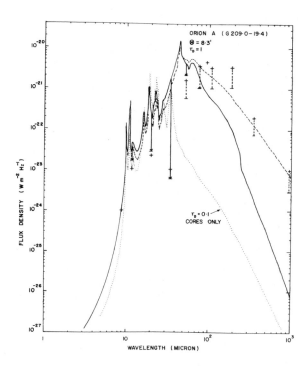

Fig. 5 Calculated spectra of olivine core-ice mantle grains with
mantle radius 0.15 μm and core radius 0.05 μm in the case of Orion
A. Solid curve is dust optical depth τ_D= 1. Dashed curve, τ_D= 1
but with a maximum enhancement of far infrared emissivities due to
impurities. Dotted curve, τ_D= 0.1 and no ice mantles on the grains.
Shown as crosses are the following observations with the beam
diameter and chopping offset as indicated: 8.6 μm (Ney et al.1973;
5", 6"), 11.6 μm and 20 μm (Ney and Allen 1969; 26", 51"), 34 μm
(Low et al. 1973; 25", 45"), 54 μm, 73 μm and 180 μm (Erickson et
al. 1973; 4', 12'), 91 μm (Harper 1974; IRel), 100 μm (Hoffmann
et al. 1971; 12', 18'), 350 μm (Harper et al. 1972; 1', 1.8'),
1000 μm (Harvey et al. 1974; 8' × 4' map). Dashed vertical lines
of the far infrared observations indicate a reduction by a factor
of three in order to subtract the contribution by the K-L Nebula.
Solid vertical lines show the predicted flux densities for the
solid curve for a beam width and a chopping offset as in the
corresponding observations.

to subtract the contribution by the K-L Nebula. The 90 μm point is the value
given by Harper (1974) for the HII region after he subtracted a point source
at the K-L central position from the total flux.

One of the difficulties in comparing observations at different wavelengths
with each other and with a theoretical spectrum is the often large range in

beam widths and chopping offsets from one measurement to another. We have
therefore used our model to calculate the predicted flux density for a beam
size and chopping offset that correspond to the observational values. The
reduction from the total flux density in the case of the solid curve is
shown by the vertical solid lines with arrows. All the corrected values are
within small factors of the observations. Clearly, with the strong and sharp
features of olivine in the 10 and 20 μm regions, the calculated values become
sensitive to small variations in the effective wavelength of the observation.
However, none of the values in Fig. 5 represent peak values, so they may not
be very different for the actual interstellar silicate material (Aannestad
1975). The strong ice peak at 45 μm may also be weaker and more diffuse if
the ice exists in its amorphous state rather than in a crystalline form.
The observations in the 50 μm region indicate somewhat smaller fluxes than
predicted for olivine core-ice mantle grains for $\tau_D = 1$. Unfortunately, with
the existence of the K-L Nebula nearby, it is very difficult to ascertain
the true flux from the HII region at far infrared wavelengths. One may even
assume that the HII region is as much as a factor of 10^3 times fainter than
the K-L Nebula at these wavelengths and still have reasonable agreement with
the shorter-wavelength observations. This is shown by the dotted curve in
Fig. 5, which assumes that the dust is clean silicate cores of 0.05 μm radius
with an effective optical depth in the Lyman continuum of just 0.1. At this
small optical depth the diffuse Lyman-α photons contribute about 50% of the
total infrared energy. However, except for the disappearance of the peak in
the intensity profiles due to the core-mantle region and slightly higher
relative intensities in the outermost region, the intensity distributions are
roughly the same as in Fig. 4. Thus, small and relatively hot silicate dust
particles may be responsible for the infrared emission from Orion A if it is
assumed that the far infrared emission is negligible compared to the near
infrared emission.

Fig. 6 shows calculated spectra for W3 for optical depths $\tau_D = 1$ and $\tau_D = 2$.
In both cases a maximum enhancement of far infrared emissivities is assumed.
The solid lines with arrows show the reduction in the predicted flux
densities at 10 and 20 μm for the case of $\tau_D = 2$. Comparison with the
observations shows a rough agreement, except that again the observations
point to less flux at 40 to 50 μm than predicted for the olivine core-ice
mantle grains.

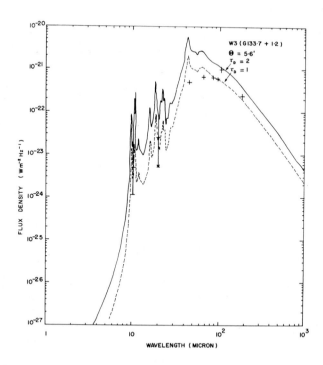

<u>Fig. 6</u>. Calculated spectra for W3. Parameters and symbols are as in Fig.5, but maximum enhancement of far infrared emissivities has been assumed. Observational points and corresponding beam widths and chopping offsets are as follows: 10 μm and 20 μm (Wynn-Williams et al. 1972; W3A + B + C; 10", 30"), 44 μm, 64 μm, 79 μm, 94 μm, and 186 μm (Harper 1974; 5', 6' - 8'), 100 μm (Hoffmann et al. 1971; 12', 18').

Fig. 7 shows the calculated spectra for τ_D = 1 compared with the observations in the case of M17. The solid curve assumes maximum far infrared impurity enhancement, and the dashed curve assumes no enhancement. The solid vertical lines take account of beam size effects, at the wavelengths of 22 and 350 μm. Note that the plotted 10 μm flux density is the integrated value from a map of ∿5' arc extent (Kleinmann 1973). However, the observed intensities are much smaller and the observed profile is much broader than in our model. The full-width at half maximum intensity at 10 μm is calculated to be about 5" arc, whereas the observations show a half-width of about 50" arc for the southern component and an even larger width for the northern component (Kleinmann 1973). Grains that are much smaller than 0.05 μm or grains that are less efficient emitters in the far infrared than silicate particles would contribute to the 10 μm intensity over a larger region. However, calculations in our model with

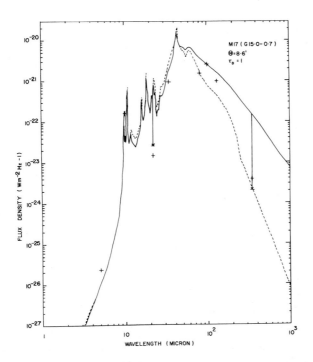

<u>Fig. 7</u> Calculated spectra for M17. Parameters and symbols are
as in Fig. 5. Solid curve: τ_D = 1 and maximum enhancement of far
infrared emissivities. Dashed curve: τ_D = 1 and no enhancement.
Observational points and corresponding beam widths and chopping
offsets are as follows: 5 μm and 22 μm (Kleinmann 1973; N + S
components, 30", 67"), 10.2 μm (Kleinmann 1973; 5' Map), 34 μm,
82 μm, and 132 μm (Olthof 1974; 30'), 100 μm (Hoffmann et al. 1971;
12', 18'), 350 μm (Rieke et al. 1973; 1.1', 4.3').

with 0.01 μm uncoated olivine particles or with 0.05 μm uncoated graphite
particles result in a half-width of at most 10" arc at 10 μm. Even for small
optical depths of $\tau_D \simeq 0.1$ in which case the diffuse Ly-α photons contribute
most of the infrared luminosity, the intensity profile at 10 μm is much
narrower than observed. The Ly-α photons may heat the grains up to ∿ 150 K
over 90% of the nebula, but most of the contribution to the 10 μm flux still
comes from hotter grains near the center of the region. Perhaps in this case
the assumption of a spherical HII region is particularly bad. Also, we have
assumed that the ionizing stars are at the exact center of the nebula. If
there are several stars and they are distributed over a volume < 1' arc
(corresponding to 0.5 pc) in diameter, the 10 μm intensity width would be

comparable to the observed width.

TABLE 2.   MODEL PARAMETERS. *

HII region	Grain sizes	τ_D	θ_0 (arc sec)	θ_{ev} (arc min)	f_D	L_{IR} ($10^5 L_0$)	x_α	x_g (10^{-12})	$\frac{M_D}{M_{HII}}$
Orion A	0.05μm cores	0.1	2	–	0.07	0.48	0.55	0.7	0.0008
Orion A	0.05μm cores; 0.15μm mantles	1	2.8	2.4	0.50	2.5	0.16	1.1	0.010
W3	"	1	1.2	1.0	0.52	17	0.15	0.7	0.007
W3	"	2	1.8	1.4	0.77	44	0.06	1.5	0.014
M17	"	1	3.2	2.6	0.50	42	0.16	0.5	0.005

* τ_D = effective dust optical depth in the Lyman continuum; θ_0 = angular diameter of inner "hole"; θ_{ev} = angular diameter of inner dust shell; f_D = fraction of the number of Lyman continuum photons absorbed by the dust; L_{IR} = total infrared luminosity; x_α = fraction of the infrared luminosity contributed by diffuse Ly-α photons; x_g = relative number density of dust to gas; M_D/M_{HII} = dust-to-hydrogen mass ratio.

Table 2 shows model parameters for some of the cases with spectra in Figs. 5 through 7.   Except for the case τ_D = 0.1 for Orion A, the values are based on calculations with a maximum far infrared emissivity, but differ insignificantly if this assumption is relaxed.

The characteristic diameter of the central "hole" in the dust distribution is a few seconds of arc. Because the temperature of the dust in the innermost region varies rapidly with distance from the center, the size of the "hole" is not sensitive to the assumed value of the evaporation temperature of the core material. The position of the inner boundary of the core-mantle region, on the other hand, depends sensitively on the evaporation temperature of the mantle material. For example, an increase in this temperature from 100 to 150 K decreases the diameters of this boundary by a factor of about two. The extent of the core-mantle region will therefore depend strongly on the chemical composition of the mantle material. Also, it should be emphasized that in a real nebula, with density variations and a distribution of grain sizes, the boundaries of the two grain regions will be considerably diffuse.

In the cases where $\tau_D = 1$, the dust absorbs about 50% of all the continuum photons. For the Orion region a single star of spectral class of about O6 would thus be luminous enough to both heat the dust and ionize the gas, consistent with θ^1 Ori C as the main exciting star in the Trapezium. For W3 and M17, however, more than one single O star may be required although the actual number depends strongly on the (unknown) spectral classes of the exciting stars. Finally, Table 2 shows that for $\tau_D \simeq 1$ the dust-to-gas number ratios and the dust-to-hydrogen mass ratios correspond to typical interstellar values and do not violate the upper limit set by the availability of heavy atoms in making up the grains.

5. Conclusions

Core-mantle grains may coexist with the ionized gas within HII regions, their distribution being limited by the evaporation of the core material in the innermost regions and by the evaporation of mantle material in an outer region. Comparison of the predicted correlation between the total infrared flux and the radio continuum flux density with observations shows that the effective dust optical depth τ_D in the Lyman continuum is of the order of unity for such a model. For $\tau_D = 1$ the dust absorbs about 50% of the total number of continuum photons, and if the continuum photon energy corresponds to O5-O6 type ionizing stars, the Ly-α photons contribute about 16% of the total infrared luminosity.

If the grains are similar to olivine core-ice mantle grains, the mantles may exist over a substantial portion of an HII region. The calculated intensity profiles show a bright inner core at 10 and 20 μm due to the core grains and a much more extended profile at far infrared wavelengths due to the core mantle grains. The observed structure changes rapidly with wavelength in the 20 to 50 μm region. However, M17 has a much more extended 10 μm intensity profile than calculated, indicating that the assumption of a spherical region with the exciting stars all at the center may not be valid in this case.

The calculated spectra for Orion A, W3, and M17 assuming olivine core-ice mantle type grains are much broader than for homogeneous type grains, and give reasonable agreement with the present near infrared and far infrared observations for optical depths τ_D = 1 to 2. Some enhancement of the infrared emissivity due to impurities seems required for $\lambda > 100$ m, whereas the calculated fluxes for "pure" grains are somewhat larger than observed for wavelengths between 40 and 60 μm. For Orion A, where all of the far infrared radiation may be ascribed to the nearby K-L Nebula, agreement with the 10 to 34 μm observations may also be obtained if the dust is uncoated silicate grains of total optical depth $\tau_D \simeq 0.1$. Except for the latter case, the dust masses needed to account for the observed infrared fluxes correspond to a normal interstellar dust-to-hydrogen mass ratio of about 0.01.

This work was supported by NSF grant MPS 73-04897. The University of Arizona kindly provided computing facilities.

References

Aannestad, P.A., 1973, in Interstellar Dust and Related Topics, IAU Symposium No.52, ed. J.M. Greenberg and H.C. van de Hulst.

Aannestad, P.A., 1975, Astrophys. J. , in press.

de Jong, T., Israel, F.P., and Tielens, A.G.G.M., 1975, preprint.

Emerson, J.P., Jennings, R.E., and Moorwood, A.F.M., 1973, Astrophys.J., 184, 401.

Erickson, E.F., Swift, C.D., Witteborn, F.C., Mord, A.J., Augason, G.C., Caroff, L.J., Kunz, L.W., and Giver, L.P., 1973, Astrophys.J., 183, 535.

Furniss, I., Jennings, R.E., and Moorwood, A.F.M., 1974, in "HII Regions and the Galactic Centre" Proceedings of Eighth ESLAB Symposium, ed. A.F.M. Moorwood, ESRO SP-105, p.61.

Greenberg, J.M., and Hong, S.S., 1974, in "HII Regions and the Galactic Centre", Proceedings of Eighth ESLAB Symposium, ed. A.F.M. Moorwood, ESRO SP-105, p.153.

Greenberg, L.T., 1973, in Interstellar Dust and Related Topics, IAU Symposium No.52, ed. J.M.Greenberg and H.C. van de Hulst, p.413.

Harper, D.A., 1974, Astrophys.J., 192, 557.

Harper, D.A., and Low, F.J., 1971, Astrophys.J. 165, L9.

Harper, D.A., Low, F.J., Rieke, G.H., and Armstrong, K.R., 1972, Astrophys. J., 177, L21.

Harvey, P.M., Gatley, I., Werner, M.W., Elias, J.H., Evans, N.J., Zuckerman, B., Morris, G., Sato, T., and Litvak, M.M., 1974, Astrophys. J. 189, L87.

Hoffman, W.F., Frederick, C.L., and Emery, R.J., 1971, Astrophys.J. 170, L89.

Isobe, S., 1970, Publ. Astr. Soc. Japan 22, 429.

Kleinmann, D.E., 1973, Astrophys. Lett. 13, 49.

Low, F.J., Rieke, G.H., and Armstrong, K.R., 1973, Astrophys. J. 183, L105.

Mathews, W.G., 1969, Astrophys. J. 157, 583.

Münch, G., and Persson, S.E., 1971, Astrophys. J. 165, 241.

Ney, E.P. and Allen, D.A., 1969, Astrophys. J. 155, L193.

Ney, E.P., Strecker, D.W., and Gehrz, R.D., 1973, Astrophys. J. 180, 809.
O'Dell, C.R., 1965, Astrophys. J. 142, 1093.
O'Dell, C.R., Hubbard, W.B., and Peimbert, M., 1966, Astrophys. J., 143, 743.
Olthof, H., 1974, Astr. Astrophys. 33, 471.
Panagia, N. 1974, Astrophys. J., 192, 221.
Petrosian, V., and Dana, R.A., 1975, Astrophys. J., 196, 733.
Pottasch, S.R. 1974, Astr. Astrophys., 30, 371.
Rieke, G.H., Harper, D.A., Low, F.J., and Armstrong, K.R., 1973,
                                        Astrophys. J., 183, L67.
Schraml, J., and Mezger, P.G., 1969, Astrophys. J. 156, 269.
Simpson, J.P., 1973, Publ.Astr. Soc. Pac. 85, 479.
Wright, E.L., 1973, Astrophys. J., 185, 569.
Wynn-Williams, C.G., Becklin, E.E., and Neugebauer, G., 1972, Mon.Not.R. Ast.
                                        Soc. 160, 1.

DISCUSSION

Bussoletti   Could you quote how much your results will be changed taking
into account realistic absorption efficiencies which must surely be less
than 1.

Aannestad We would need the detailed behaviour of the absorption efficiency
of dust particles in the Lyman continuum to estimate the change. This is not
currently available either from laboratory work or astronomical methods, but
I do not expect the results to be significantly different.

POLYFORMALDEHYDE GRAINS

A. Cooke and N. C. Wickramasinghe

Department of Applied Mathematics and Astronomy,

University College, Cardiff, Wales

Abstract   Arguments are presented for the occurrence of
polyformaldehyde mantles on interstellar grains.

Astronomers have for some while been accustomed to considering two types of
grains[1] :-

		Melting Point
(i) Refractory grains:		
(e.g. Graphite, silicates, etc.)		1500 - 2000 K
(ii) Volatile grains:		
(e.g. ices, H_2O etc.)		100 - 200 K

The occurrence of grains of intermediate volatility with melting

temperatures $\sim$ 400 - 500 K typified by crystalline polymers (e.g. polyacetals,

polyformaldehydes) have not been considered so far.

Refractory grains are most likely to form in conditions of relatively high

density and temperature prevailing in the atmospheres of giant stars or in

explosive situations such as in novae[2] or supernovae[3]. In C-rich situations

a predominance of C-grains is expected, in oxygen-rich cases silicates. The

composition of mantles which could condense in interstellar clouds is more

difficult to assert with confidence. Ices of various types have been

considered to date, but the case of polymerization and the formation of

mantles of polyformaldehydes appears to be equally strong. The most likely

venue for the formation of such mantles is in molecular clouds where significant gas phase abundances of H_2CO are already observed. There is observational evidence of a very close correlation between the dust and H_2CO distribution in these clouds. Whilst H_2CO molecules are always strongly correlated with dust, OH is not.

Three steps are usually required for polymerization:
(1) Initiation.   This involves the conversion of an

$$
\begin{array}{ccc}
H & & H \\
O = C & \text{structure to} & O - C - \\
H & & H
\end{array}
$$

and could be achieved by the action of a UV photon or a radical or an ion.

(2) The propagation of the chain. This would proceed spontaneously, since it is an exothermic process.

(3) End-capping, and termination of chains. Reactions with either water or methanol, both of which are present, could produce this effect. H_2O could form $H(CH_2O)_nOH$ whilst CH_3OH produces $CH_3(CH_2O)_nOH$ - the former being stable up to temperatures $\sim$ 150 K and the latter to 450 K under ambient interstellar conditions.

The spectral features of POM (polyformaldehyde) at 3.4μ, 10μ, 18μ and their probable astronomical identification have already been discussed in a preliminary way[4,5].   We have now obtained $n(\lambda)$, $k(\lambda)$ from a Kramers-Kronig analysis of the absorbance data of polyformaldehyde films heated to various temperatures. Using this data we can now calculate $Q_{abs}(\lambda)$ and hence emission spectra for small spherical grains. Figure 1 shows the function

$$
F(\lambda) \propto Q_{abs}(\lambda) \; B(\lambda, T)
$$

with T = 445 K, in the vicinity of the 10 μm band together with a normalised emission spectrum of the Trapezium Nebula.   The agreement with astronomical data such as the emission excess in the Trapezium nebula is generally satisfactory. Preliminary calculations for silicate core - POM mantle grains indicate that closer agreement is possible for this case.   A detailed account of these calculations will be published elsewhere[6].   The 10 and 18 μm emission bands observed in regions such as the Trapezium nebula could thus

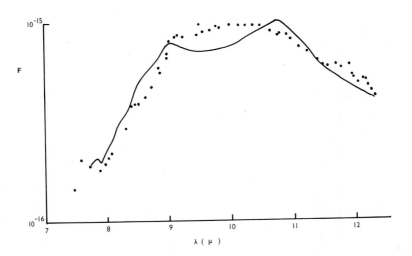

Fig. 1 $Q_{abs}(\lambda) B (\lambda, T), T = 445$ K, for POM grains compared with
a normalised emission spectrum of the Trapezium nebula.

arise from one or other or a combination of at least two types of solid
material.

The existence of materials typified by polyformaldehyde in cometary dust is
strongly indicated by a recent analysis[7] of the 10 and 18 μm bands in comets
Bennett and Kohoutek.   Both these bands and the underlying infrared
continuum are depressed by a factor 10 in intensity a few days after
perihelion where the radiative temperature is only ∿ 500 K. Whilst this data
is consistent with the evaporation of polyformaldehyde, it is inconsistent
with pure silicate grains which can evaporate only at temperatures > 1400 K.
Since cometary dust is likely to have an interstellar origin, there is strong
circumstantial evidence for the existence of interstellar polyformaldehyde.
This conclusion is further supported by the actual detection of
polyformaldehyde in the Allende carbonaceous chondrite[8].

References

1. Wickramasinghe, N.C., 1967, Interstellar Grains (Chapman & Hall, London)
2. Clayton, D.D., and Hoyle, F., 1975, Astrophys.J. (in press)
3. Hoyle, F. and Wickramasinghe, N.C., 1970, Nature, 266, 62.
4. Wickramasinghe, N.C., 1975, Mon.Not.R.Astr.Soc., 170, 11P.

5. Wickramasinghe, N.C., 1974, Nature, 252, 462.
6. Cooke, A., and Wickramasinghe, N.C., 1975, in preparation.
7. Mendis, D.A., and Wickramasinghe, N.C., 1975, Astrophys.Sp.Sc.
                                                      in press.
8. Breger, I.A., Zubovic, P., Chandler, J.C. and Clarke, R.S., 1972,
                                    Nature 236, 155.

DISCUSSION

Joseph What is the effect of the proposed polymeric grain composition on
the extinction curve across the entire spectrum into the ultraviolet?
Wickramasinghe The long-chain polymers would tend to grow in the form of
dielectric whiskers. Whiskers of radii 10^{-5} cm could explain the available
data on interstellar extinction and polarization in the waveband 3500 $\overset{o}{A}$-1 μm.
The extinction and albedo data in the far ultraviolet requires a comparable
mass of smaller uncoated graphite and silicate grains.
Rengarajan How good is the association between dust clouds, especially those
showing the 10 μm features, and H_2CO emission?
Wickramasinghe The association between extinction, infrared emission (and
also millimeter wave radiation) on the one hand and H_2CO contours on the
other is very close for molecular clouds (e.g. Sag A). The OH correlation
with dust is not so close.
Silk   How do you reconcile the observed gas phase abundance of formaldehyde
with the polymer density required by your model?
Wickramasinghe The value of the gas phase H_2CO density could be understood
in terms of a polymer-monomer equilibrium situation. The observed gaseous
H_2CO is what is knocked off due to the effect of ultraviolet and thermal
evaporation. The idea that there is a substantial solid-state store of H_2CO
(formed in protostellar situations and ejected into interstellar space)
is an attractive one because the direct formation of H_2CO in normal tenuous
interstellar clouds presents many problems.
Bussoletti Can you distinguish, by mass arguments, between formaldehyde
polymers and silicates in explaining the infrared excess of the Trapezium
nebula?
Wickramasinghe The mass estimates are comparable for both materials.
Melchiorri Have you data about the spectral features of these polymers
in the far infrared between 20 μm and about 300 μm? If absorption lines
exist in this region, this fact may be used to distinguish between polymers
and silicates.
Wickramasinghe All the data I have seems to stop at about 50 μm in the far
infrared. Besides the 10 and 18 μm bands there is another fairly strong
feature at 45 μm.

SUPERGRAIN MODELS OF FAR INFRARED SOURCES

M. G. Edmunds and N. C. Wickramasinghe

Department of Applied Mathematics and Astronomy,

University College, P.O. Box 78, Cardiff, Wales

Abstract A new model for infrared and far-infrared sources is proposed. The use of thin whisker grains of lengths up to 1 mm, perhaps multiply branched to form "snowflake"-like particles, allows much greater far-infrared emission than is possible with an equivalent mass of conventional small spherical grains. The growth of such whiskers and their application to models of Galactic and extragalactic sources is discussed.

A wide variety of sources have been observed whose radiation output shows a peak in the infrared at $\lambda \sim 120$ μm, with remarkably similar spectra. The sources include regions of active star formation (e.g. the Orion molecular cloud), galactic nuclei (particularly Seyfert nuclei) and possibly quasi-stellar objects (QSO). Possible explanations for the 100μ peak which have been proposed for extragalactic sources include (1) processes occurring around small accreting primordial black holes (2) Non-linear inverse Compton losses from electrons in a multiple pulsar model and (3) molecular maser lines smeared out by Doppler broadening. After these rather exotic processes, the suggestion of re-radiation of a primary source by dust (4) seems almost conventional! For Galactic sources, dust models seem to be nearly universally accepted.

A problem arises from the amount of dust required to give the observed fluxes in the far-infrared, due to the rapid fall-off in emission efficiencies for small grains at long wavelengths. Taking very rough values for the infrared fluxes over a 1 mm bandwidth at 1 mm wavelength of 3×10^{41} erg s^{-1} for a Seyfert galaxy and $10^{44} - 10^{46}$ erg s^{-1} for a QSO, and using the optimum theoretical emission efficiencies for small spherical grains, implies dust masses of $>5 \times 10^6$ $M_\odot$ for Seyferts and $> 10^9 - 10^{11}$ $M_\odot$ for QSO's. The details of these and subsequent calculations may be found in (5) and (6). The dust masses must be regarded as excessive, particularly considering both

the localisation required for a Seyfert source and the fact that we have taken
the theoretical maximum absorption efficiency which will not be met for real
grains.   Although Galactic sources give reasonable dust masses ($\sim$ 1M$_\Theta$ for
the Orion Molecular cloud), there are problems with the high surface
brightness (7).   We propose that the far-infrared emission can be more easily
explained by having a contribution from dust particles in the form of highly
elongated cylinders or "whiskers".   Using the relevant infinite cylinder
formula for grains of length $\ell \gtrsim \lambda$, and allowing metallic grain lengths
comparable with the millimetre wavelengths gives rise to dust masses (5), (6)
of 4 $\times 10^3$ M$_\Theta$ for Seyferts, $10^6 - 10^8$ M$_\Theta$ for QSO's.   The Orion mass could be
reduced as far as 0.02 M$_\Theta$, with no surface brightness problems.

Is whisker growth unreasonable? We are not considering the bulk of the
interstellar medium but only region of high density c.f. molecular clouds.
Obviously there would exist a range of shapes and sizes of grains, giving rise
to intermediate mass estimates. A large number of elements are observed to
grow as whiskers under laboratory conditions – e.g. iron, or graphite growth
from screw dislocations. If we allow dielectric materials (although this will
increase somewhat the grain mass required), then simple polymerised organic
molecules (e.g. formaldehyde polymers) with dirty ice mantles would be a good
candidate. The usual assumption that grains will grow as spheres is perhaps
a rather arbitrary one, and there is a strong need for laboratory experiments
on the morphology of crystal growth at low temperatures and pressures.

One great advantage of whisker growth is its rapidity. Provided the
diffusion lengths of accreted atoms are longer, or comparable with the length
of the growing whisker, then the whole grain acts as a collecting area for
growth at its tip. Thus the rate of growth is exponential provided there is
no exhaustion of gas phase atoms. For example, a whisker of radius 10^{-6}
could grow to a length of 1 mm in only 10^6 years in a typical protostellar
environment.   This rapidity of growth is a considerable advantage over the
very long timescales required (on the basis of simple accretion) for the very
large spherical grains envisaged by Rowan-Robinson (7). In order to avoid
depletion of gas phase atoms before long whisker growth has occurred, the
number of growth sites must be much less than the canonical gas/dust number
density. This could perhaps be achieved by sputtering of initial grains, or
by only limited initial nucleation.

In conclusion, whisker grains can enhance infrared and far-infrared emissivities and may be expected to grow very rapidly under suitable conditions. The whiskers may be multiply branched in the form of "snow-flakes", but assuming some asymmetry remains we would expect high far-infrared polarisation from sources, provided an alignment mrechanism exists. Observational study of variability of far-infrared emission remains a strong test of this, and amy other, dust model for extragalactic sources.

Acknowledgment

We are very grateful to Dr. M. Rowan-Robinson for stimulating discussion.

References

(1) Mezaros, P., 1975, Private communication.
(2) Arons, J., Kulsrud, R.M., and Ostriker, J.P., 1975, Astrophys.J. 198, 687.
(3) Solomon, P.M., and Rees, M.J., 1972, Mem.Roy.Soc.Liege, 6th series, 3,591.
(4) Rees, M.J., Silk, J.I., Werner, M.W., and Wickramasinghe, N.C., 1969,
              Nature 223, 788.
(5) Edmunds, M.G., and Wickramasinghe, N.C., 1975, Nature in press.
(6) Edmunds, M.G., and Wickramasinghe, N.C., 1975, Mon.Not.R.Astr.Soc.
              Submitted.
(7) Rowan-Robinson, M., 1975, Mon.Not.R.Astr.Soc. 172, 109.

DISCUSSION

Greenberg Did you arrive at a reduced mass estimate for elongated particles compared with spheres using the relative emissivity (10 a/λ) for spheres? If so, you may note that in this case the effective ratio should be significantly reduced because you have already overestimated the sphere efficiency factor.
Edmunds The emission efficiencies of metallic cylindrical grains were calculated using the rigorous formula for cylinders (Wickramasinghe, N.C., Light Scattering Functions for Small Particles with Applications in Astronomy, Adam Hilger Press, London, 1973). I would mention similar formulae derived by J.M. Greenberg and G. A. Shah (Astr.Astrophys. 12, 250, 1971), particularly page 252, with the requirement that we are considering highly elongated cylinders.
Williams (1) Will not grain-grain collisions with these long grains cause break-up of grains into short stubs? (2) Even if you got a long whisker grain, will it not tend to wind up into a ball and so simply resemble a spherical grain of equivalent mass?
Edmunds (1) Collisions would cause some break-up, but remember that the number density of these long grains will be much less than the usual small sphere number densities, and hence collisions will be less frequent. (2) In laboratory conditions whiskers seem to grow fairly straight. Obviously there will be some branching, and we would expect a much larger and open structure than for a spherical grain of equal mass.

THEORETICAL MODELS OF DUST CLOUDS

M. Rowan-Robinson

Queen Mary College, London E.1

Abstract Models of optically thick dust clouds are fitted to the spectra of Galactic (Orion and W3) and extragalactic (M82 and NGC253) sources from 3 µm to 1 mm. If the grains are composed of ice or silicates, then radii greater than 10 µm are required. These giant grains appear to be a general feature of sources peaking in the far infrared.

Dust models of far infrared sources associated with dense molecular clouds and compact HII regions have been proposed by several authors (Greenberg 1971, Pottasch 1974, Bollea & Cavaliere 1975, Rowan-Robinson 1975a). These models are self-consistent only for the optically thin case, and where heating of the grains in the outer part of the cloud by interior grains can be neglected. Although these sources are thin at wavelengths $\gtrsim$ 100 µm, they are optically thick at near infrared and optical wavelengths, and integration of the radiative transfer equation in the non-grey, optically thick case is necessary. The interesting physics of the ionized gas and molecules can only be properly treated in the framework of a correct picture of the energy flow through the cloud. Judging by the fact that the main energy output from these clouds is in the form of continuum radiation with a quasi-thermal spectrum peaking in the range 30 - 300 µm, this energy flux is controlled by the dust. Rather a small fraction of the energy from the illuminating source or sources goes into maintaining the ionization of the gas or exciting molecules, and this has been neglected in the present work.

Spherically symmetric, grey, optically thick clouds have been studied in the context of circumstellar dust shells (e.g. Huang 1969, Herbig 1970, Schwartz 1975). An attack on the non-grey case, neglecting scattering, was made by Larson (1969), and Rowan-Robinson (1975b) applied a model of this type to the Orion nebula, with a more realistic assumption about the absorption efficiency of the grains. That work is still not strictly self-consistent since the

adopted temperature distributions do not ensure flux conservation through
the cloud. Self-consistency is the main innovation of the work reported here.

Let $T(r)$, $n(r)$ be the temperature and number-density of grains at distance
r from the centre of the cloud, and suppose that

$$n(r) = n_2(r/r_2)^{-\beta}, \quad r_1 \leq r \leq r_2,$$

and is zero otherwise. Then the luminosity of the cloud per steradian due to
the dust is (Rowan-Robinson 1975b)

$$P(\nu) = 4\pi \int_{r_1}^{r_2} r^2 G_\nu(r) B_\nu(T(r)) \alpha_\nu \, dr \tag{1}$$

where

$$G_\nu(r) = \frac{1}{2r} \int_0^r \frac{s \, ds}{(r^2-s^2)^{1/2}} \{ \exp(-\int_{\max(s,r_1)}^{r_2} f(y) dy$$

$$- \int_{\max(s,r_1)}^r f(y) dy + \exp(-\int_r^{r_2} f(y) dy) \}$$

$$f(y) = \alpha_\nu y (y^2 - s^2)^{-1/2}$$

$$\alpha_\nu = \pi a^2 Q_\nu n(r)$$

$$a = \text{the radius of the grains (assumed spherical)}$$

$$Q_\nu = \text{the absorption efficiency of the grains}$$

$$B_\nu(T) = \text{the Planck function}$$

and it is assumed that the grains radiate like blackbodies weighted by the
factor Q_ν.

At frequencies at which the cloud is optically thin, $G_\nu(r) \simeq 1$, and if also $h\nu \ll kT_{min}$, where $T_{min} = T(r_1)$, then

$$P(\nu) = \frac{8\pi k T_{min} A r_2^2 q_\nu \nu^2}{(3 - \beta - \delta)c^2} \qquad (2)$$

where $T(r) \propto r^{-\delta}$, $A = \pi a^2 Q_{uv} n_2 r_2$, $q_\nu = Q_\nu/Q_{uv}$.

Figure 1 shows the total extinction efficiency for different grain materials compared with the smoothed Q_ν adopted in the present work (see Rowan-Robinson 1975b) for different grain radii

$$Q_\nu = 10 \ x/\lambda(\mu m), \qquad max(12.5 \ \mu m, 0.8\pi a) < \lambda$$

$$= 0.8 \ x, \qquad 12.5 \ \mu m < \lambda < 0.8\pi a$$

$$= 2, \qquad \lambda < 0.8\pi a \qquad (3)$$

where $x = 2\pi a/\lambda$.

At long wavelengths this will be a reasonable approximation for both silicates and ice. At short wavelengths (3) will correctly allow for total extinction, but will overestimate emission by a factor of 2.

It is found that the frequency of the peak in $P(\nu)$ is determined almost entirely by T_{min}. The far infrared part of the spectrum then determines, using (2) and (3), the quantity Aa. Table 1 shows values of T_{min} and Aa determined in this way for Orion, W3, M82 and NGC253, assuming the values for r_2 shown.

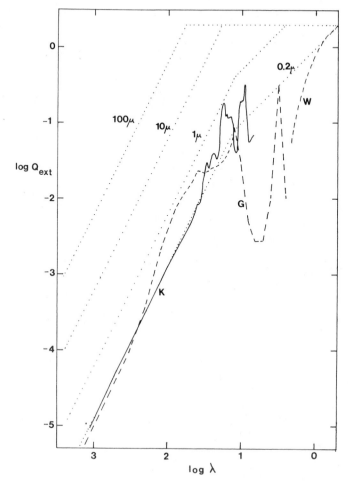

<u>Fig. 1</u>. Extinction efficiency, Q_ν, for 0.2 μm silicate (K, Knacke & Thomson 1973, sample 10058) and H_2O ice (G, Gaustad 1963, absorption only) grains, and for graphite cores, radius 0.07 μm, with ice mantles, radius 0.21 μm (W, Wickramasinghe 1967). The dotted curves show the Q_ν adopted in the present work for different grain radii.

The mass-density in grains is then

$$\rho_{gr} \;\; = \;\; \frac{4 A a \bar{\rho} (1 \, - \, (r_1/r_2)^{3-\beta})}{Q_{uv} r_2 (3 \, - \, \beta)} \tag{4}$$

TABLE 1.

	assumed angular diameter	distance	r_2	T_{min} (K)	Aa(m)	n_{H_2} (cm^{-3})	M_{gr} ($M_\odot$)
Orion	4' [1]	0.5 kpc	0.29 pc	40	300	1.7×10^6	85
W3	100" [2]	3 kpc	0.73 pc	40	300	6.6×10^5	527
M82	25" [3]	3.1 Mpc	188 pc	30	125	1.07×10^3	1.46×10^7
NGC253	18" [4]	3.4 Mpc	148 pc	30	125	1.37×10^3	9.1×10^6

[1]Adopted to conform to beams used for 2-200 m data (see Fig. 2), [2]Fazio et al 1974, [3]Hargrave 1974, [4]Becklin et al 1973.

and assuming a dust to gas ratio of 1% by mass, that the average density of the grain material $\bar{\rho}$ = 2.5 gm cm^{-3}, and that the cloud is uniform, we obtain the values shown for the number-density of hydrogen molecules n_{H_2} and the total mass in grains M_{gr}.

The value of n_{H_2} for the Orion cloud, 1.7×10^6 cm^{-3} is inconsistent with the value deduced by Goldsmith et al (1975) from their observations of the J = 1 - 0 and 2 - 1 transitions of ^{12}CO and ^{13}CO, 2×10^3, but is consistent with the values $10^5 - 10^6$ deduced from H_2CO (Harvey et al 1974) and DCN (Phillips et al 1974), and the value $\geq 2 \times 10^5$ given by Liszt et al (1974) from their analysis of CO and CS.

To fit the spectra of the sources we have to find a temperature distribution which will preserve flux constancy through the cloud, i.e.

$$\int_0^\infty (P^{(o)}(\nu, R) - P^{(i)}(\nu, R))d\nu = \int_0^\infty P_c(\nu)\{1 - \exp(-\int_{r_1}^R \alpha_\nu dr)\}d\nu \quad (5)$$

for $r_1 \leq R \leq r_2$, where $P_c(\nu)$ is the luminosity of the central illuminating source per ster, $4\pi P^{(o)}(\nu, R)$ is the total outward flux across the surface $r = R$, $4\pi P^{(i)}(\nu, R)$ is the total inward flux, given by (1) with

$$G_\nu^{(o)}(r) = \frac{1}{2r} \int_0^r \frac{s\ ds}{(r^2-s^2)^{1/2}} \{\exp(- \int_{\max(s,r_1)}^R f(y)dy - \int_{\max(s,r_1)}^r f(y)dy)$$

$$+ \exp - \int_r^R f(y)dy\} \qquad \text{for} \qquad r_1 < r < R$$

$$= \frac{1}{2r} \int_0^R \frac{s\ ds}{(r^2-s^2)^{1/2}} \{\exp(- \int_{\max(s,r_1)}^R f(y)dy - \int_{\max(s,r_1)}^r f(y)dy)\}$$

$$\text{for} \qquad R < r < r_2 \qquad\qquad (6)$$

$$G^{(i)}(r) = \frac{1}{2r} \int_0^R \frac{s\ ds}{(r^2-s^2)^{1/2}} \exp(- \int_R^r f(y)dy) \qquad \text{respectively.}$$

If the central source emits most of its energy at wavelengths such that Q_ν = constant, then the R.H.S. of (5) can be written

$$\frac{\{1 - \exp(- \int_{r_1}^R \alpha_\nu\ dr)\} \int_0^\infty P^{(o)}(\nu, r_2)d\nu}{1 - \exp(- \int_{r_1}^{r_2} \alpha_\nu dr)} \quad .$$

Models satisfying (5) to an accuracy of a few percent for $r_1 \leq R \leq r_2$ have been found with a uniform density distribution ($\beta = 0$) and

$$T = T_{max}(r/r_1)^{-\delta_1}, \qquad \text{for} \qquad 100\ K < T < T_{max}$$

$$= T_{min}(r/r_2)^{-\delta_2}, \qquad \text{for} \qquad T_{min} < T < 100\ K \qquad\qquad (7)$$

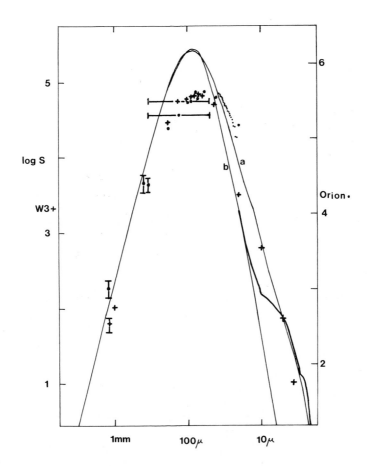

Fig. 2. Spectra of Orion (dots and heavy line, RH scale) and W3 (crosses, LH scale) compared with theoretical models (a) and (b) - see text. References for Orion data: 1.23 mm, Clegg et al 1975*; 408 μm, Soifer & Hudson 1974*; 350 μm, Gezari et al 1974*; 40 - 350 μm, Emerson et al 1973; 42, 59, 72, 78, 91, 99, 183 μm, Harper 1974; 20 - 40 μm, Houck et al 1974; 20 μm, Lemke et al 1974; 2.2 - 10 μm (heavy line), compilation by Gatley et al 1974. (* corrected to diameter 4' arc, centred on BN). W3 data: 1.23 mm, Clegg et al 1975 (combined flux from W3A & B); 1 mm and 2 - 20 μm, Wynn-Williams et al 1972; 40 - 350 μm, Emerson et al 1973; 44, 64, 79, 94, 186 μm, Harper 1974.

where $T_{max} = T(r_2)$ and has been taken as 1000 K.

The spectra of Orion and W3 (Fig. 2) have been compared with two such models:

(a) A = 1, a = 300 μm, δ_1 = 0.45, δ_2 = 0.5, (b) A = 4, a = 75 μm, δ_1 = 0.35,
δ_2 = 0.5. Both agree reasonably with the distribution of T_{CO} for Orion
over the range r = 1 - 3' arc. Bearing in mind that the sources are in
fact complex, with two or more distinct condensations, that probably not all
the radiation originating from the central sources has been included, that
the Orion spectrum includes the contribution of the Trapezium nebula, that
the theoretical curves include the contributions neither of the attenuated
central source nor of scattered light, and the many other simplifications
of this treatment, the fit is reasonable. If the grain radius was of order
0.1 μm, as required to explain ultraviolet and visual extinction properties,
the ultraviolet optical depth would be of order 1000 and that at 10 μm
would be of order 100. The observed 1 - 10 μm radiation in these sources,
and indeed in most such dense clouds, is far too strong relative to the
100 μm - 1 mm radiation for this to be the case. A coincidental geometry,
which places the near infrared sources near the front of the dust cloud in
the line of sight (Gezari et al 1974, Zuckerman & Palmer 1974) could be invoked
if we had only one or two sources with spectra of the type shown in Fig. 2:
however these spectra are fairly typical. Figure 4 compares the intensity
distribution across the cloud predicted by model (a) with the strip scan in
Right Ascension across the BN source in Orion by Clegg et al 1975. The
theoretical curves are labelled with the assumed value of the angular radius.
That with r = 3' arc is a reasonable fit bearing in mind that the data is
convolved with the telescope beam profile and has been smoothed. In the
optically thin part of the spectrum the intensity distribution given by
these models falls off approximately exponentially with distance except near
the centre and edge.

The far infrared spectra of the galaxies NGC253 and M82 have been compared
(Fig. 3) with model (c) A = 1, a = 125 μm, δ_1 = 0.5, δ_2 = 0.4. Again the fit
is reasonable, allowing for a contribution in the near infrared from stars,
and the possibility of small grains alone being present is ruled out. The
optical depth at 10 μm, $\sim$ 1, agrees well with that inferred by Gillett et al
(1975) for these two galaxies from 8 - 13 μm observations of silicate
absorption. The fact that HII regions and optical condensations are visible
in the optical nuclei of these galaxies is entirely consistent with this
model.

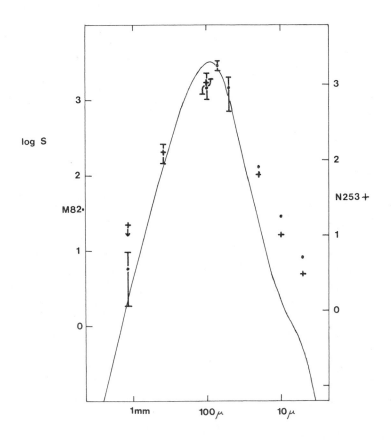

<u>Fig. 3</u>. Spectra of M82 (dots, LH scale) and NGC253 (crosses, RH scale) compared with theoretical model (c) - see text. Observational data: 1.23 mm, Ade et al 1975; 400 µm, Rieke et al 1973; 47, 65, 100 µm, Harper & Low 1973; 2 - 20 µm, Rieke & Low 1972 and Becklin et al 1973.

In all cases considered here appreciable masses of small ($\sim$ 0.1 µm) grains could be present without making any significant contribution to the far infrared radiation in these sources. Either a continuous distribution of particle sizes, as in the solar system, or a bimodal distribution, are appropriate. Inclusion of the effect of scattering is likely to bring down somewhat the required grain sizes, so the values quoted here should not be taken too

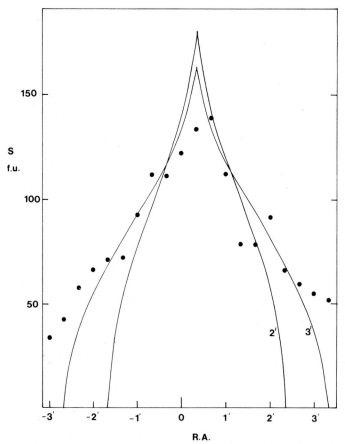

<u>Fig. 4.</u>   Intensity distribution predicted by model (a) assuming
r = 2 & 3 ' arc, compared with strip scan in R.A.   across the
Orion BN   source by Clegg et al 1975.

literally. The formation of the large grains presumably takes place, as in
the solar system, during the dense final stages of star formation.

The simplest alternative to the giant grain hypothesis is to find a grain
material with an absorption efficiency in the far infrared a factor of 100
times greater than silicates or ice. This is in effect the proposal of
Andriesse & Olthof (1973), who postulate

$$Q = (\lambda_o/\lambda)^2, \quad \lambda > \lambda_o,$$

$$= 1, \quad \lambda < \lambda_o, \quad \text{with} \quad \lambda_o \sim 30 \ \mu m.$$

(It can be seen from Fig. 1 that the appropriate value of λ_o for silicates and ice is ~ 3 μm.) But this material has yet to be identified.

References

Ade, P.A.R., Rowan-Robinson, M. & Clegg, P.E., 1975, to be published.
Andriesse, C.D. & Olthof, H., 1973, Astr.Astrophys., 27, 319.
Becklin, E.E., Fomalont, E.B. & Neugebauer, G., 1973, Astrophys.J., 181, L27.
Bollea, D. & Cavaliere, A., 1975, preprint.
Clegg, P.E., Ade, P.A.R. & Rowan-Robinson, M., 1975, to be published.
Emerson, J.P., Jennings, R.E. & Moorwood, A.F.M., 1973, Astrophys.J., 184, 401.
Fazio, G.G., Kleinmann, D.E., Noyes, R.W., Wright, E.L., Zeilik, M. & Low, F.J., 1974, Astrophys.J., 192, L23.
Gatley, I., Becklin, E.E., Matthews, K., Neugebauer, G., Penston, M.V. & Scoville, N., 1974, Astrophys.J., 191, L121.
Gaustad, J.E., 1963, Astrophys.J., 138, 1050.
Gezari, D.Y., Joyce, R.R., Righini, G. & Simon, M., 1974, Astrophys.J., 191, L33.
Gillett, F.C., Kleinmann, D.E., Wright, E.L. & Capps, R.W., 1975, Astrophys.J., 198, L65.
Goldsmith, P.F., Plambeck, R.L. & Chiao, R.Y., 1975, Astrophys.J., 196, L39.
Greenberg, J.M., 1971, Astr.Astrophys., 12, 240.
Hargrave, P.J., 1974, Mon.Not.R.astr.Soc., 168, 491.
Harper, D.A., 1974, Astrophys.J., 192, 557.
Harper, D.A. & Low, F.J., 1973, Astrophys.J., 182, L89.
Harvey, P.M., Gatley, I., Werner, M.W., Elias, J.H., Evans, N.J., Zuckerman, B., Morris, G., Sato, T. & Litvak, M.M., 1974, Astrophys.J., 189, L87.
Herbig, G.H., 1970, Astrophys.J., 162, 557.
Houck, J.R., Schaack, D.F. & Read, R.A., 1974, Astrophys.J., 193, L139.
Huang, S.S., 1969, Astrophys.J., 157, 843.
Knacke, R.F. & Thomson, R.K., 1973, Publ.astr.Soc.Pacific, 85, 341.
Larson, R.B., 1969, Mon.Not.R.astr.Soc., 145, 297.
Lemke, D., Low, F.J. & Thum, C., 1974, Astr.Astrophys., 32, 231.
Liszt, H.S., Wilson, R.W., Penzias, A.A., Jefferts, K.B., Wannier, P.G. & Solomon, P.M., 1974, Astrophys.J., 190, 557.
Phillips, T.G., Jefferts, K.B. & Wannier, P.G., 1974, Astrophys.J., 192, L153.
Pottasch, S.R., 1974, Astr.Astrophys., 30, 371.
Rieke, G.H., Harper, D.A., Low, F.J. & Armstrong, K.R., 1973, Astrophys.J., 183, L87.
Rieke, G.H. & Low, F.J., 1972, Astrophys.J., 176, L95.
Rowan-Robinson, M., 1975a, Proc. 2nd European meeting of I.A.U., Trieste, ed. E. Müller.
Rowan-Robinson, M., 1975b, Mon.Not.R.astr.Soc., 172, 109.
Schwartz, R.D., 1975, Astrophys.J., 196, 745.
Soifer, B.T. & Hudson, H.S., 1974, Astrophys.J., 191, L83.
Wickramasinghe, N.C., 1967, Interstellar Grains (Chapman & Hall, London).
Wynn-Williams, C.G. & Becklin, E.E., 1974, Publ.astr.Soc.Pacific, 86, 5.
Zuckerman, B. & Palmer, P., 1974, Ann.Rev.Astr.Astrophys., 12, 279.

DISCUSSION

Emerson With such large grains would not the far infrared emission be optically

thick? And do the observations not suggest that the far infrared emission
is optically thin?

Rowan-Robinson My cloud models go thick at wavelengths between 50 and
90 μm and are optically thin at longer wavelengths. Since the product Aa
is kept constant, an increase in the grain radius a is accompanied by a
decrease in the number-density, leaving the optical depth at say 1 mm
unchanged.

Aannestad The emission efficiency of ice in the far infrared varies more
like ν^3 than ν^2 from about 100 μm onwards. How would this change
your values for the source properties?

Rowan-Robinson It would mean that even larger grains would be needed (and
even larger masses in grains), if they were composed of ice.

Whitworth Have you considered having sources of excitation distributed through
the source and/or clumps in the radiation-transfering medium? You might be
able to avoid invoking such a drastic grain model.

Rowan-Robinson Both these effects would be difficult to deal with mathematic-
ally. If the extent of the cluster of exciting sources were small compared
with the overall radius of the cloud, as seems to be implied by the strongly
peaked spatial distribution of the far infrared radiation, then this would
not change things much. I would have to think about the effect of clumps.

Edmunds Are you not worried by the large growth times for such large grains?
Even though the earth is good evidence that large grains can form under
certain circumstances, the time-scales for growth of 100 μm spheres on the
basis of classical theory are much longer than the typical evolution time-
scales of star formation.

Rowan-Robinson Let me put it this way: if such grains exist outside the
solar system, how would we learn about them except through their far infrared
radiation? Such grains obviously do form in planetary systems, if not more
generally, but I admit that I do not have a theory for getting them into
these dense molecular clouds.

Gilra I want to ask whether it is possible to have an observational check on
the large grains you are considering. One could in principle observe hydrogen
recombination lines in the radio and observe the Balmer lines, and then
determine the total extinction at, say, H_β. With this information it may
be possible to take out any contribution due to large particles - there will
not be any reddening in the optical region due to large particles.

Rowan-Robinson No, and I think the neutral extinction of these large·grains
in the visible and ultraviolet would be hard to detect, since it would
probably be a rather small fraction of the total extinction.

Andriesse It is disturbing that the size of solid particles invoked to explain
data in a certain wavelength range are always of the same order as the wavelength. This is true in the far ultraviolet, for the visual, now for the
far infrared and yesterday even for the millimetre background. There seem
to be traps hidden in the various arguments and I would ask all participants
to look for methods of escape from this curious rule.

Rowan-Robinson Of course in the solar system we see a spectrum of particle
sizes from 0.1 μm up to planetary size, with comparable contributions to
the mass density from each decade of size. Perhaps we should think of the
light in these dense clouds cascading from high to low frequencies, being
processed by progressively larger grains.

Wannier I take to task your very high central value of $n_{H2} = 1.7 \times 10^6$ in
Orion A. Contrary to your implication, the observed line intensities from
millimetre transitions of molecules with high dipole moments do not imply
very high gas densities. Line formation calculations including photon-trapping
imply that even these molecules are consistent with hydrogen number-densities
of $10^4 - 10^5$. This lower gas density would seem to give rise to some

embarrassment when compared to your mass of 85 $M_\odot$ in the form of dust in the central 2' arc in Orion.

Rowan-Robinson I would agree that the molecular hydrogen density I deduce for Orion is on the high side, by almost an order of magnitude, compared with even the highest values quoted from molecular line studies. I should emphasize that the actual numbers I gave should not be taken too literally in view of the large number of simplifying assumptions. But perhaps the molecular studies are not final and definitive either.

INFRARED EMISSION FROM GRAINS WITH FLUCTUATING TEMPERATURES

J. M. Greenberg

Hughens Laboratory, Leiden 2405, Holland

Abstract   The far infrared radiation by the small ($\sim$0.005 μm) grains in the bimodal interstellar size distribution is shown to deviate substantially from that predicted by steady state temperature predictions. In dark clouds it is shown that the radiation corresponds to lower temperature than for the classical ($\sim$ 0.1 μm) grains while the opposite is to be expected from hot clouds.

1. Introduction

The temperature of grains in interstellar space has classically been calculated on the basis of a steady state condition of radiative balance between emission and absorption. Since the absorption occurs largely at a wavelength comparable with the size of the classical particles and the emission by the grains occurs at wavelengths large compared with the particle dimensions, the absorption and emission efficiencies of the particle are different. Normally, the absorptivity (emissivity) decreases with increasing wavelength for a given particle size so that under standard interstellar conditions and in dark clouds, small grains tend to have higher temperatures than large grains (Greenberg 1968, 1971). The natural assumption is that, if the interstellar grain population consists of two populations of grains, one of which is of the order of 0.1 μm and the other of which is $\sim$0.005 μm, that the radiation emitted by the small grains will be shifted to shorter wavelengths relative to that emitted by the classical sized grains. However, this assumption turns out to be incorrect because the small grains can not be characterized by a steady state temperature but rather are subjected to substantial temporal fluctuations in temperature by virtue of their small steady state energy content compared with the absorbed photon energies (Greenberg 1968, Greenberg and Hong 1974).

In this paper we shall present a sample calculation of the effects of
temperature fluctuations on the emission by very small particles and on the
total radiation emitted by the combination of the normal and small particles
in interstellar space.

2. The Bimodal Distribution

The wavelength dependences of interstellar extinction and polarization in
combination with a number of other basic criteria such as the abundances of
the elements in the interstellar medium have been used to arrive at a two
population description of the interstellar grains (Greenberg and Hong 1974).
One population consists of grains with silicate cores ~ 0.05 μm radius
surrounded by modified ice mantles whose outer radius is ~ 0.1 μm; the other
population consists of much smaller particles, possibly of silicate
composition (although this is not critical in the qualitative aspects of this
paper) whose sizes are ~ 0.005 μm. The mean properties of the interstellar
grains are well represented by cylindrical core-mantle particles with silicate
cores of 0.05 μm radius and mean mantle size of 0.12 μm radius and for each
such particle an additional 4.6×10^3 e spherical bare particles with radius
~ 0.005 μm, where e is the elongation (ratio of length to width) of the core-
mantle particles. In this paper, the grain temperature will be based on these
parameters but, for simplification, the emission properties will be calculated
always for spherical grains.

3.   Steady State Temperatures

The equation of steady state absorption (from the radiation field) and
emission (by the grains) is

$$\int_0^\infty Q_{abs}(\lambda,a)R(\lambda)d\lambda = \int_0^\infty Q_{abs}(\lambda,a)B_\lambda(T_g)d\lambda \tag{1}$$

where $Q_{abs}(\lambda,a)$ is the absorption (emission)efficiency of the particle of
size a at the wavelength λ, $R(\lambda)$ is the local radiation field and $B_\lambda(T_g)$
is the Planck radiation function at the steady state grain temperature T_g.

Numerical solutions of Eq.(1) for the given core-mantle and bare particles
yield steady state temperatures

$$T_{c-m} = 15 \text{ K}$$
$$T_b = 20 \text{ K}$$

and, as anticipated, the smaller particles appear to have the higher
temperature.

4. Temperature Fluctuation

For the core-mantle grains the temperature fluctuations are essentially non-
existent. For the small bare particles, on the other hand, Table 1 shows
that a significant number of photons in the interstellar radiation field will
raise the grain internal energy to states significantly higher than the mean
temperature would imply.

TABLE 1. PEAK FLUCTUATING TEMPERATURES OF SMALL SILICATE GRAINS WITH
          a = 0.005 μm RESULTING FROM SINGLE PHOTON ABSORPTIONS.
                      Steady state temperature = 20 K

ΔU (eV)	0.1	0.5	1.0	5.0	10.0	photon energy
T_f ($^{\circ}$K)	30	45	54	64	80	fluctuating peak temperature

An indication of the kind of time dependence of the temperature is given in
terms of the characteristic times for: photon absorption, τ_{abs}, diffusion of
absorbed energy within grain, τ_{diff}, and cooling by radiation, τ_{cool}.

The mean ultraviolet photon absorption time for $\lambda \lesssim 2000$ Å is

$$\tau_{abs} \approx 10^3 \text{ sec.}$$

The time for conversion of the absorbed energy to phonon modes may be estimated as the order of the size divided by the speed of sound in the medium (neglecting surface effects) as

$$\tau_{diff} \sim \frac{2a}{V_s} \sim 10^{-11} \text{ sec.}$$

The cooling time is roughly the energy content ΔU divided by the rate of emission at the temperature implied by ΔU.

$$\tau_{cool} \sim \frac{\Delta U}{(T_{\Delta U})} \underset{\sim}{\sim} 1 \text{ sec for } \Delta U = 1 \text{ ev.}$$

We see that for the above conditions the temperature spikes are distinctly separated. Each time a photon is absorbed, the temperature rises almost instantaneously to its peak value, then relatively quickly drops to something below the average and the grain spends most of the time at a temperature below the average temperature. This is schematically illustrated in Fig. 1.

The temperature distribution function, D(T)dT, defined as the fraction of time the grain spends between T and T-dT has been calculated on the basis of random photon collisions and using the fact that $\tau_{abs} \ll \tau_{diff}, \tau_{cool}$ (Hong 1975) to be of the form shown in Fig.2 where certain approximations have been made for the average absorption properties of the grain in the far-infrared. We note again that D(T) is large for $T < T_{av}$. The

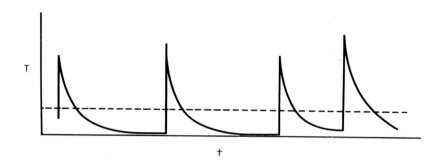

<u>Fig. 1</u> Fluctuating temperature of small grains is schematically represented as a function of time. The dashed line represents the steady state grain temperature.

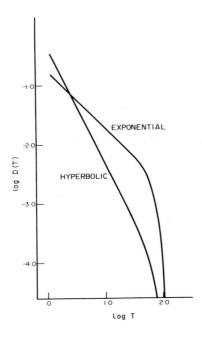

<u>Fig. 2</u> Temperature distribution function D(T). The upper curve
is for exponential cooling and the lower for hyperbolic cooling.

assumption of hyperbolic cooling corresponds to a grain emissivity $\sim \lambda^{-1}$.
Exponential cooling occurs when the emissivity over the region of importance
is independent of λ.

<u>5. Far Infrared Emission</u>

The flux from N_{c-m} core-mantle particles at temperature T_{c-m} is

$$F_{\lambda}^{c-m} = N_{c-m} 4\pi a_m^2 \pi B_\lambda (T_{c-m}) Q(\lambda, a_m) \ . \tag{2}$$

The flux from $N_b = \alpha N_{c-m}$ bare particles (α = ratio of number of bare to
number of core-mantle particles) is either

$$F_{\lambda}^{b} = \alpha N_{c-m} 4\pi a_b^2 \pi B_\lambda (T_b) Q(\lambda, a_b) \ , \tag{3}$$

or

$$F_\lambda^b = \alpha N_{c-m} 4\pi a_b^2 \pi <B_\lambda(T)> Q(\lambda, a_b) \ , \tag{3'}$$

where the bracket denotes an average over the temperature distribution function, namely

$$<B_\lambda(T)> = \int B_\lambda(T)D(T)dT. \tag{4}$$

Adding Eq.2 to Either Eq. 3 or 3', we have for the total flux from all the grains in the interstellar mixture

$$F_\lambda = N_{c-m} 4\pi a_m^2 Q(\lambda, a_m) \pi B_\lambda(T_{c-m})\{ \ 1+ (a_b/a_m)^2 \frac{Q(\lambda,a_b)}{Q(\lambda,a_m)} \frac{\exp(1.44/\lambda T_{c-m})-1}{\exp(1.44/\lambda T_b) \ -1}\} \tag{5}$$

or

$$F_\lambda = N_{c-m} 4\pi a_m^2 Q(\lambda, a_m) \pi B\lambda(T_{c-m})\{ \ 1+\alpha(a_b/a_m)^2 \frac{Q(\lambda,a_b)}{Q(\lambda,a_m)} \frac{\exp(-1.44/\lambda T_{c-m})-1}{<\exp(-1.44/\lambda T)-1>}\} \tag{5'}$$

where < > again denotes the average taken over D(T).

From the fact that, for a black body, the maximum of the emission for a temperature T occurs at a wavelength λ_m = 0.29/T, we see that for a grain temperature of ∿15 K the important spectral region is around 200 μm. For all interstellar particle sizes $2\pi a/\lambda$ is much less than unity at such wavelengths, so that the Rayleigh approximation may be applied to obtain the efficiency factors $Q(\lambda, a_b)$ and $Q(\lambda, a_m)$. For simplicity we shall do all our calculations for spheres, including the core-mantle particles. No significant difference can be attributed to the use of non-spherical particles, unless they are extremely elongated (Greenberg 1975).

The fluxes given in Eqs. (5) and (5') have been calculated by us numerically for the appropriate grain models and conditions. In this paper we shall present only results of calculations based on simplifying assumptions, which illustrate clearly the effects of temperature fluctuations on the radiation without involving detailed optical properties of grains.

In the Rayleigh approximation the ratio $Q(\lambda,a_b)/Q(\lambda,a_m)$ is proportional to a_b/a_m with equality resulting when the bare and core-mantle particles are of the same material. We shall let this proportionality be wavelength independent in order to avoid the complications introduced by the fact that the bare and core-mantle particles are of a different composition. In this case the quantity $\alpha(a_b/a_m)^2(Q(\lambda,a_b)/Q(\lambda,a_m))$ becomes the volume ratio, $\alpha(a_b/a_m)^3$, of bare to core-mantle particles, which is about 0.22 for the average interstellar grain composition (Hong 1975). Furthermore, we have ignored the wavelength dependence of the first factor $Q(\lambda,a_m)$ in Eqs.(5) and (5') because of the lack of reliable laboratory measurements of optical constants in the far infrared.

The qualitative nature of the changes introduced in the far infrared

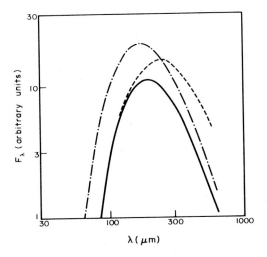

Fig. 3 Infrared radiation emitted by a bimodal dust distribution. The solid curve corresponds to core-mantle particles alone. The dash-dot curve shows the effect of adding the radiation by the bare particles assuming they are at a constant temperature. The dashed curve represents the total radiation when the small bare particles are properly considered to have fluctuating temperatures with a distribution based on hyperbolic cooling. A small difference would result from using the D(T) for exponential cooling.

radiation by the small grains is shown in Fig.3. We see that if we had
naively used the steady state temperature for the small grains, the
radiation maximum for emission by all the grains would have been predicted at
about 170 μm. However, the maximum is shifted to about 250 μm as a result of
the use of the fluctuating temperature distribution for the small grains.
The effective temperatures of the dust based on observations with such values
of λ_m would be 17 K and 12 K respectively. In addition, it is of
considerable interest to note that the total infrared radiation emitted by
the small grains is substantially larger than that by the classical sized
grains. This is a consequence of the fact that the total area of the small
grains is several times larger than that of the classical grains. Since they
intercept more of the visible and ultraviolet radiation from the interstellar
radiation field, they must re-emit a larger amount in the infrared.

In HII regions it may no longer be true that $\tau_{abs} \ll \tau_{cool}$. For example
if we increase the ultraviolet energy density by only a factor of 10^3
relative to that in the average interstellar space, the value of τ_{abs} becomes
of the order of τ_{cool}. Such a condition prevails as far as ∿5 pc from an
O5 star. In this case not only is the ratio of the temperature of small
grains to the classical grains higher than in dark clouds, but the small
grains emit even larger relative amounts of infrared. Furthermore, since
their temperature may be considered as constant, they shift the radiation
toward shorter wavelengths rather than toward longer wavelengths.

Acknowledgment

This work was supported in part by NASA grant NGR-33-011-043 and by NSF
grant MPS74-21138.

References

Greenberg, J.M., 1968, in Stars and Stellar Systems, Vol 7, Nebulae and
        Interstellar Matter, ed. by B.M.Middlehurst and L.H.Aller (Chicago:
        Univ. of Chicago Press), 221.
Greenberg, J.M., 1971, Astron & Astrophys., 12, 240.
Greenberg, J.M., 1975, in preparation.
Greenberg, J.M., and Hong, S.S., 1974, in IAU Symp. No.60, Galactic Radio
        Astronomy, ed. by F.J.Kerr and S.C. Simonson, III (Dordrecht: Reidel),
        155.
Hong, S.S., 1975, Thesis, State University of New York at Albany.

DISCUSSION

Wickramasinghe Can you give any supporting arguments for 0.005 μm grains?
Greenberg The principal arguments in favour of grains in the 0.005 μm radius
range is the curvature of the interstellar extinction curve up to 1000 Å.
The acceptable interstellar grain materials are absorbing in this region and
consequently the existence of a negative curvature at 1000 Å implies that at
this wavelength the particles must be of a size such that $2\pi a/\lambda \lesssim \frac{1}{2}$
(at $2\pi a/\lambda \simeq 1$ saturation would already be setting in). Applying this
criterion we get a $\lesssim 0.008$ μm, and when performing our detailed Mie theory
calculation we found a ~ 0.005 m to give a better representation of the
observed extinction curve in the far ultraviolet. (see Greenberg, J.M., 1973,
Interstellar Dust & Related Topics p.3; Greenberg, J.M. & Hong, S.S., 1974
HII Regions & the Galactic Centre ed.A.F.M.Moorwood, p.221; Greenberg, J.M.
& Hong, S.S., 1974, IAU Symposium No.60, Galactic Radio Astronomy, ed. F.J.
Kerr and J.C.Simonson (Reidel), p.155).
Wickramasinghe Does the Mie theory apply to such small particles?
Greenberg Mie theory certainly applies to grains of this size because the
bulk optical properties - certainly of dielectrics - are valid for this
amount of material. However, Mie theory does not apply if the particles are
irregular, particularly if they have surface irregularities. Experimental
results obtained in my microwave laboratory in the Laboratory for Particle
Scattering, Albany, show that highly absorbing particles with irregularities
produce extinction significantly different from equivalent spheres when the
irregularities are of order 15% (see Greenberg, J.M., Wang, R.T. and Bangs,
L., 1971, Nature Phys.Sci. 230, 110).
Wickramasinghe Are low energy cosmic-ray protons likely to destroy these
grains?
Greenberg If there were low energy protons of significant number in the
cosmic-ray spectrum, they would destroy grains. However, there is no
evidence for such a low energy tail in the c.r. spectrum and, furthermore, its
original raison d'etre, namely the production of the intercloud medium, seems
no longer to exist.
van Duinen Are these small particles stable against the energy and
temperature fluctuations?
Greenberg The small particles are stable against the energy and temperature
fluctuations, which are too small to break them. However, the temperature
fluctuations are very probably the reason why the small particles cannot
accrete dirty ice mantles in the way larger (0.05 μm) cores do.

THE PROTOSTELLAR ORIGIN OF INTERSTELLAR GRAINS

J. Silk

Department of Astronomy, University of California,

Berkeley, Calif. 94720, U.S.A.

Abstract A semi-quantitative description of the origin of interstellar grains in a protostellar environment is given. Grains
grow in the protostellar collapse phase and can be expelled by
radiation pressure from stars of mass $\gtrsim 5\ M_\odot$. A convective shell
of grains forms in which shattering produces a power-law spectrum
of fragments. A bimodal distribution of grain sizes, peaking at
$\sim 0.1\ \mu m$ and $\sim 0.01\ \mu m$ results. The effects of protostellar
winds and rotation are considered.

1. Introduction

Far infrared astronomy essentially consists of the observation of emission
and absorption by dust grains in interstellar or circumstellar regions. To
understand the origin of these particles is therefore an important aim that
may shed light on their physical properties. In this paper I shall describe
some recent work on this problem (Burke & Silk 1975) that provides a semi-
quantitative description of the origin of interstellar grains in a protostellar environment.

Herbig (1969) suggested that stars of $\sim 1 M_\odot$ could in their early protostellar
phases provide a prolific source of grains by ejection of small particles from
a protoplanetary accretion disk by means of a protostellar wind. Here I shall
propose an alternative mechanism, namely that in stars more massive than
$\sim 5 M_\odot$ the protostellar radiation suffices to expel a significant fraction of
the accreting material. I shall also consider the effects of protostellar
T-Tauri-like winds on the infalling material, and will provide a more
quantitative discussion of Herbig's hypothesis. One of the principal aims of
my discussion is to emphasize the role of various physical processes (such as

309

shattering, fusion, accretion and nucleation) in determining the size
distribution of the ejected grains. Tentative conclusions that are reached
include the attainment of a <u>bimodal</u> grain size distribution, and an
explanation of the origin of interstellar grains.

Recent evidence that supports a protostellar origin for interstellar grains
includes observations of interstellar depletion, which tend to support a
proportionality between condensation temperature and depletion (Field 1974,
Morton 1974), and the apparent increase in grain size in dark clouds
(Carrasco et al. 1973). It is therefore of interest to first discuss the
collapse of an interstellar cloud.

2. Infall and Grain Shell

Consider a dense interstellar cloud that becomes gravitationally unstable to
collapse and fragmentation. The manner in which this occurs has been
extensively discussed in the astrophysical literature, and it suffices for
the present purpose to consider a fragment of an initially massive cold cloud
that is just Jeans unstable, so that the radius of the fragment is

$$R_J = 0.5 \ (M_{10}/T_{10}) \ pc,$$

where M_{10} denotes its mass in units of $10 \ M_\odot$ and T_{10} the gas temperature
in units of $10 \ K$.

Larson (1973) has given a comprehensive survey of the collapse and evolution
of a collapsing cloud. The collapse tends to be extremely non-homologous,
with the formation of a dense central core. The behaviour in the outer
regions can be approximated by a similarity solution appropriate to isothermal
fluid flow with constant velocity $u \simeq 3 \ v_s$, where v_s is the sound speed,
and a density profile of the form $\rho = A \ r^{-2}$, where r is the distance from
the centre of the cloud and A is a constant to be evaluated from the
initial conditions. The outward pressure gradient produced as a consequence
of the core formation decelerates the collapse considerably from the initial
free fall. The temperature remains at $\sim 10 \ K$ until the central core

becomes opaque to the infrared radiation by the dust grains, and heats up
and accretes more matter to become a protostar. For protostars in excess of
$\sim$ 5 $M_\odot$, a substantial luminosity is produced while a considerable fraction
of the initial cloud is still accreting.

During the early phases of the collapse, one can readily show that the
heavier atoms and molecules will be almost completely depleted onto the
surfaces of dust grains. Fusion of the smaller grains by thermal collisions
will also occur at sufficiently high densities: one finds that in the case
of a strongly peaked grain size distribution, grains of radii $\stackrel{<}{\sim} 0.05$ μm
that are outside the evaporation region at $\sim$ 10 A.U. (see Eq. (1)) will
have fused.

The adiabatic central core develops after approximately one free-fall time of
the initial configuration. Matter falling onto this core encounters a shock
front as it decelerates, resulting in emission of a substantial luminosity.
A 5 $M_\odot$ protostar attains an initial luminosity of $\sim$ 10^3 $L_\odot$, which lasts
for $\sim$ 10^6 yr, comparable to the initial collapse time. Protostellar models
with $M \stackrel{>}{\sim} 5$ $M_\odot$ attain their peak luminosity while substantial accretion of
the initial cloud is still occurring, and the present discussion primarily
concerns stars in this mass range.

Radiation from the central object will be completely absorbed by the infalling
matter. As a consequence of the large optical depth, the momentum of the
protostellar luminosity is deposited in a thin shell which interacts
dynamically with the infalling material. The inner radius of this shell,
determined by the temperature T_g at which the solid particles evaporate,
is given by

$$r_e = 5 \times 10^{13} (2000 \text{ K}/T_g)^{5/2} (L/10^3 L_o)^{\frac{1}{2}} (Q_v/0.3) \text{ cm} \qquad (1)$$

where Q_v is the mean extinction efficiency of a grain and L is the
protostellar luminosity. The effective shell thickness is determined by the
scattering, absorption, and reemission of the ambient radiation, and solution
of the radiative transfer problem indicates that after approximately 100
optical depths, the grain temperature becomes independent of the incident

radiation field, maintaining a temperature of ~ 10 K in the case of graphite grains (Werner & Salpeter 1969). This corresponds to a shell thickness

$$X \cong 4 \times 10^{12}(a/10^{-5} \text{ cm})(r_e/10^{14} \text{ cm})^2 \text{ cm}, \qquad (2)$$

beyond which one can approximate the infalling material as isothermal. Radiation pressure acts on the grains only within the first few optical depths of the shell, and the situation of interest is one in which the radiation pressure can overcome the opposing forces of gravity and ram pressure, and cause the grain shell to move outward. Individual grains that are newly exposed to the radiation field r_e are rapidly decelerated and forced outward beyond a few optical depths of material, where infalling material drags them in again. The grain shell will escape from the central protostar if the protostellar luminosity exceeds a certain critical value, found, by considering the dynamical evolution of the grain shell ejected through an isothermal infalling medium (Burke & Silk 1975), to reduce to

$$L \gtrsim 3000 \ T_{10}^{2} \ L_{\odot} \ . \qquad (3)$$

Here T_{10} is the initial temperature of the gas cloud in units of 10 K, assumed to be originally at its Jeans mass. The protostellar luminosity is required to exceed this value over a time-scale

$$t \gtrsim 10^5 (L/10^3 \ L_{\odot})^{-\frac{1}{2}}((R_J - u \ t)/10^{18} \text{ cm}) \text{ yr}. \qquad (4)$$

The grain shell accretes the bulk of the infalling material as it moves beyond r_e. The lowest mass protostar which satisfies conditions (3) and (4) has approximately 5 $M_{\odot}$, and recent models (Westbrook & Tarter 1975) indicate that approximately 10% of the initial Jeans mass is in the outer envelope when the core luminosity peaks. For a 20 $M_{\odot}$ protostar, this fraction is about 60%, and radiation pressure limits the maximum protostellar mass to ~ 50 $M_{\odot}$ (Larson & Starrfield 1971, Kahn 1974).

It has already been remarked that the effect of the radiation pressure on the
innermost grains will be a source of turbulent motions in the innermost part
of the shell. In fact, the entire shell may be shown to be convectively
unstable. Although the bulk of the protostellar luminosity is absorbed and
reradiated by grains, a fraction ($\sim$ 1%) of the incident energy is expended
in heating the gas by grain-atom collisions. Standard mixing length arguments
yield an estimate of the mean convective velocity to be

$$v \cong 2\ T_{10}^{1/6}(a/0.1\ \mu m)^{1/3}(r/10^{14}\ cm)^{2/15}(L/10^3 L_o)^{1/15}\ km\ s^{-1}\ .$$

The shell is also Rayleigh-Taylor unstable; however the maximum scale of this
instability is similar to the convective scale-height, and it is probably
that both convection and the effects of transverse pressure gradients act to
suppress the Rayleigh-Taylor modes.

3. Grain Size Distribution

The convective motions in the grain shell have an important effect on the
grains, because grain-grain collisions are relatively frequent at the high
ambient density. Since colliding grains will possess relative kinetic energies
comparable to their chemical binding energies, grain collisions will be
capable of disrupting individual grains. Fusion can be shown to be relatively
unimportant compared to shattering.

The distribution of fragment sizes from the shattering of a solid particle of
radius a can be taken to be

$$n(a, a') = \theta\ a^{\varepsilon}/(a')^{1+\varepsilon}\ ,$$

where a' is the radius of the fragment, θ is a mass normalization factor,
and ε is a parameter ($1.5 \lesssim \varepsilon \lesssim 3$) which is fitted to experimental data.
Experimental evidence provides support for a power-law dependence on fragment
size down to $a' \lesssim 10\ \mu m$ (Hartmann 1969); moreover particles in the earth's

atmosphere exhibit a power-law spectrum down to sizes as small as ~ 0.03 μm (Junge 1963).

The requirement that the energy expended in breaking lattice bonds by the increased surface area of the fragments as compared to that of the original grain yields a lower limit on the size of grains which can undergo shattering (if $\varepsilon < 2$) of

$$a_{min} = 6 E_b \frac{3 - \varepsilon}{2 - \varepsilon} (1 - f)^{2/3} (\rho_d v^2 \sigma_b)^{-1} \quad ,$$

where E_b is the energy per lattice band, σ_b is the area per bond, f is the mass fraction vaporized per collision, and ρ_d is the density of grain material. For $\varepsilon > 2$, the expression for a_{min} depends on the minimum fragment size, but yields a similar result if $\varepsilon \lesssim 2.5$. One find that for typical refractory grain materials, $a_{min} \simeq 0.1$ μm.

In addition to fragmentation, grain growth can occur by accretion of ambient gas atoms. Moreover at the inner edge of the grain shell, the gas may be supersaturated with grain-forming monomers, and nucleation of a substantial number of very small grains can occur. Application of quasi-equilibrium nucleation theory (Burke & Silk 1975) indicates a critical nucleation particle size (defined to be the maximum size for which evaporation is just balanced by accretion of monomers) of approximately 30 monomers. Nucleation occurs when the grain shell first moves outwards in roughly the bottom 10% of the shell thickness.

The following scenario emerges for the evolution of the grain size distribution. The bulk of the grains are likely to be initially larger than a_{min}. The Protostellar luminosity for $M \gtrsim 5$ $M_\odot$ will drive a convective shell outward, and the convective motions will lead to shattering of the grains. The vaporized monomers will, in the inner region of the shell, provide a source for nucleation of extremely small solid particles. Continuing infall will also supply grains, and accretion of monomers will enable grains to grow. If sufficient growth occurs, grains may become large enough to shatter again, and continue this process.

An important source of uncertainty in making a quantitative calculation of the resulting grain size distribution is the fraction f of the material vaporized in a typical shattering collision. If, as seems reasonable, $f \sim 0.5$, one finds that the size distribution of the shattered fragments initially increases in mean size, but stays below a_{min}. This tends to inhibit shattering, reducing the supply of monomers, and decreases the shift in size. If f is small, a pure shattering spectrum results; if f is close to unity, a relatively flat size distribution is found.

The net effect of all this is that a _bimodal_ size distribution results, consisting of early fragmentation products that grow to just below a_{min} (~ 0.1 μm), and therefore fail to shatter again, and later fragments that do not grow appreciably and peak at the minimum fragment size a_{frag} (~ 0.01 μm). In addition the nucleation products may be comparable to a_{frag}.

4. Protostellar Winds and Rotation

Mass loss via a protostellar wind may also provide an important dynamical interaction with the infalling material. Although the interpretation of T-Tauri stars is not unambiguous, there is considerable evidence for mass outflow from these systems which are still in the evolutionary stage of pre-main sequence contraction, at velocities of several hundred kilometers per second, the inferred mass loss rates amounting to $\sim 10^{-7}$ $M_{\odot}$ yr^{-1} (Kuhi 1964). Such a wind can form a dense shell behind the shock front where it interacts with the infalling material, and, if strong enough, could eject the shell. This process would again be important for more massive stars, since the wind must be produced while the initial cloud is still collapsing.

The dynamical evolution of the shell is similar to that of the radiation driven shell. One finds now that shell ejection occurs for winds of velocity v with mass loss rates

$$\dot{M} \gtrsim 5 \times 10^{-8} (100 \text{ km s}^{-1}/v) T_{10} \ M_{\odot} \ yr^{-1} \quad .$$

A significant difference with the radiation driven shell arises in the
internal structure of the shell. For a given momentum flux, a wind expends
less energy than radiation by a factor v/c. Consequently, convection is
found to be relatively insignificant for a wind-driven shell. Although
shattering does not therefore play any role, the emergent grain distribution
is nevertheless bimodal, consisting of small, nucleated particles and large
grains that emerge relatively unscathed.

Rotation of the protostellar nebula may be of particular importance for the
less massive stars in enabling the protostar to retain or accrete an
extended, flattened, centrifugally-supported circumstellar nebula. Such a
nebula is likely to be unstable to fragmentation, and a conventional scenario
is that a thin disc will form of solid particles that range up to planetary
dimensions. Continuing accretion of gas and disruptive collisions of
planetesimals should provide a continuing source of small particles, although
their orbital lifetime may be limited by the Poynting-Robertson effect. A
relatively modest protostellar or stellar wind will suffice to eject the
smaller particles from the system, the criterion being that particles of radii

$$a \lesssim 0.3(\dot{M}/10^{-10} \; M_\odot \; yr^{-1})(v/100 \; km \; s^{-1})(M/M_\odot)\mu m$$

will be ejected.

5. Astrophysical Implications

The principal aim of this paper has been to attempt to account for the origin
of the interstellar grains. Therefore, it is pertinent to estimate the
contributions provided by stars in various mass ranges.

The mean density in the form of grains in the Galaxy is $\rho_{dust} \sim 10^{-4} \; M_\odot \; pc^{-3}$;
adopting a mean grain lifetime of $\sim 10^9$ yr, the required grain formation
rate is $\sim 10^{-13} \; M_\odot \; pc^{-3} \; yr^{-1}$. Consider first stars of mass below $\sim 5 \; M_\odot$.
Here one resorts to a protoplanetary disk and a protostellar wind to provide
ejection of small particles. Herbig (1969) argued that for stars like the
sun, if the mass initially in small particles was comparable to the mass in

the planets $(1.4 \times 10^{-3} M_\odot)$, one could obtain a substantial contribution
to the mass of interstellar grains by this mechanism. If there are ~ 0.02
stars pc^{-3} that each contribute $\sim 10^{-3} M_\odot$ in small particles, there
would result $\sim 20\%$ of the required mass in interstellar grains. Evidently
the considerable uncertainties in this estimate make it exceedingly imprecise.

In the case of stars in excess of $\sim 5 M_\odot$, one can now be somewhat more
definitive. Here one can be fairly certain that ejection of a radiation-
driven grain shell occurs. Integration over a mass-function appropriate to
stars between 5 and 50 $M_\odot$, and utilizing protostellar model calculations
to estimate the mass in the outer regions of the collapsing nebula when the
protostellar luminosity peaks, yields an ejection rate of $\sim 4 \times 10^{-14} M_\odot pc^{-3}$
yr^{-1}. This comes close to the required value.

If indeed the more massive stars provide the dominant source of interstellar
grains, some important inferences can be drawn about the grain size
distribution. Most of the mass resides in the larger grains. The nucleated
grains and the smaller shattered fragments contribute negligibly to the mass
and mean optical cross-section because of their small size. Most of the
optical effect is produced by the largest shattering fragments (with sizes
$\sim a_{min} (1 - f)^{1/3}$), although the small fragments greatly _outnumber_ the larger
ones. If $\varepsilon > 2$, the surface area is dominated by the smaller grains: this
may be of importance in connection with rates of depletion or molecule
formation.

The equilibrium spectrum of grains in the interstellar medium will be determined
by a balance between grain growth by accretion in dense clouds, injection
of new grains, and grain destruction processes. Thus by increasing the
number of grains, one has a natural means of recycling grains by mixing
with undepleted matter, and increasing the mass in grains after successive
generations of star formation. The primordial grains may have been nucleated
in a protostellar environment, and can subsequently grow and multipy by
protostellar 'procreation'.

Grains in diffusive clouds will not accrete as significantly, and the bimodal
size distribution implies that the relatively large number of small particles
may be able to account for the far ultraviolet extinction law. Apparent

spatial variations in the ratio of far ultraviolet excess to the   2200 $\overset{o}{A}$ peak
may also be understood in terms of variations in the size distribution of the
smaller particles in different regions.

Acknowledgements

This paper was written during a brief visit to the Institute of Astronomy,
Cambridge. This research has been supported in part by the Afred P. Sloan
Foundation and by NASA under grant NGR 05-003-578.

References

Burke J.R. & Silk J., 1975, in preparation.
Carrasco L., Strom S.E. & Strom K.M., 1973, Astrophys.J. 182, 95.
Field G.B., 1974, Astrophys.J. 187, 453.
Hartmann W.K., 1969, Icarus 10, 201.
Herbig G.H., 1969, Contrib.Lick Obs. No.302.
Junge C.E., 1963, Air Chemistry and Radioactivity, Academic Press, p.117.
Kahn F.D., 1974, Astron.Astrophys. 37, 149.
Kuhi L.V., 1964, Astrophys.J. 140, 1409.
Larson R.B., 1973, Fundamentals of Cosmic Physics 1, 1.
Larson R.B., & Starrfield S., 1971, Astron.Astrophys. 13, 190.
Morton D.C., 1974, Astrophys.J. 193, L35.
Werner M.W. & Salpeter E.E., 1969, Mon.Not.R.Astr.Soc. 145, 249.
Westbrook R. & Tarter C.B., 1975, Astrophys.J. (in press).

DISCUSSION

Whitworth Can you elaborate this statement that you actually increase the
amount of grains - rather than just reconstituting them? Since it appears
that at the onset of star formation a very appreciable fraction of the heavy
elements are already in grains, you cannot increase the amount - rather you
are just reestablishing the initial grain size distribution.
Silk If grains are present initially in more or less the standard amount,
the total mass in grains is not affected by this mechanism (other than by
depletion in the dense collapsing cloud phase), but the number of grains can
be greatly increased. Hence when this grain material is diluted with interstellar matter, and eventually undergoes a further cycle of protostellar
processing, the grain nuclei will provide the means of accreting additional
particles, thereby increasing the mass in grains.
However if no grains are present to begin with, as one might expect to be the
case early in the history of the Galaxy, nucleation in a protostellar
environment can provide the initial grain nuclei, and successive recycling
through the protostellar environment will produce the observed amount of
grains.
Andriesse To determine the lower limit to the size of 0.1 µm you need to
compare the kinetic energy of the fragments with the surface energy, so you
have to put in values for the bond strengths. Will the size limit not be
influenced by the strengths you put in and would you not as easily get a
value of 1 µm or 0.01 µm?

Silk For refractory materials (silicates, graphite etc), the range in bond
strengths yields an estimate of the uncertainty in the minimum size for
shattering of rather less than a factor of 10.

Williams I believe Gahm argues that the T-Tauri mass loss is in fact a
misinterpretation of observation, though I personally am going to make use
of the wind in the next talk.

Silk I believe that there is observational evidence in the T-Tauri phase
for both infall and outflow at velocities in excess of the stellar escape
velocity. Although the conventional picture has been to assume that outflow
predominates the form of an energetic wind, I would agree that the actual
situation must be rather more complicated.

DUST IN PROTOPLANETARY SYSTEMS

I. P. Williams

Queen Mary College, London E.1

Abstract It is of general interest to determine whether other
planetary systems exist and by what mechanism they came to be
formed. An outline is given of current thinking on planetary
formation, outlining the important part played by dust grains
in the process.

One of the major problems facing anyone attempting to formulate theories
for the origin of the planetary system is the fact that we know of the
existence of only our own. The difficulty here is that one cannot
therefore distinguish between what has come into being in our system as a
consequence of some fortuitous event and what might reasonably be expected
to be present whenever planetary systems are formed. This, of course,
accounts for the rough correlation that is to be found between the number of
theories and the number of astronomers.

Detecting planetary systems other than our own is therefore a work of some
importance and considerable effort has been devoted to this, most notably by
van de Kamp (e.g. 1971) using astrometric means. Infrared observations
could also play their part here since the ratio of Jupiter to solar brightness in the infrared is much more favourable than in the visible. There is,
however, some difficulty with resolution, and it is difficult to be certain
that one has two distinct sources of radiation.

What I wish to suggest here is that it may be far easier to detect protoplanetary systems, especially in the infrared, because of the large amount of
dust that is almost certainly present at such an epoch. Success in such an
operation may have a double benefit, not only confirming the existence of
other planetary systems, but also giving evidence as to the mode of formation.

Since I am not an observational astronomer, what I propose to do is to give an account of some of the main steps in the later stages of the formation of the planets, at least according to current scientific thought.   Further details can be found in reviews such as ter Haar and Cameron (1963), Williams and Cremin (1968), Woolfson (1969) and Williams (1974).   I do not intend to refer to theories of specific authors, but rather to work backwards from the existing system and deduce a set of alternatives which must have existed.

It is sensible to start with some relevant facts concerning our planetary system.   Broadly speaking, the planets can be divided into three types, differing from each other in mass, chemical composition and position relative to the Sun.   These are: the terrestrial planets typified by the Earth, the major planets by Jupiter and the outer planets by Neptune.   The quantities of interest are given in Table 1.

TABLE 1

Planet	Mass (Earth Masses)	Composition	Relative Abundances
EARTH	1	Iron-Silicate	.003
JUPITER	318	Solar	1.00
NEPTUNE	17	C,N,O based compounds	.05

It is immediately noticeable that the mass divided by the relative abundance gives in each case a value very close to 300.   In other words, one of the following states must have existed.

A)   At one stage there was a system of identical objects, each of mass roughly comparable to that of Jupiter and of solar composition.   Near the Sun only the iron-silicate type grains within such objects survived while at large distances only the icy grains have survived.   The intermediate objects kept all their mass.   Detailed calculations suggesting how this may have come about are given by McCrea and Williams (1965), Williams and Crampin (1971), Williams and Handbury (1974), Handbury and Williams (1975).

The simple picture is that the grains present in such an object segregate

under gravity towards the centre of the object, growing in the process.
Close to the Sun, these grains are of the iron-silicate variety and hence the
core which forms is Earth-like. The proximity of the Sun, with its
associated radiation, tidal effect and solar wind, results in the gaseous
layers being driven off, leaving only the core.   Further out, these effects
diminish and the gas is not removed from the major planets.   At even larger
distances the temperature falls to a level where substances like ammonia and
methane freeze onto the grains. Hence about 5% of the mass is now
segregating and the subsequent release of energy is sufficient to evaporate
the hydrogen layers, leaving a Neptune-like object.

The problem of interest here is the detection of such blobs of dust and gas.
Each blob has a surface area of the order of $10^{25} cm^2$, so that all ten of them
represent an area of $10^{26} cm^2$.   If we observe from out of the plane of the
system so that we see radiation re-emitted (or scattered) by the grains, it
seems that the infrared radiation emitted would be far easier to detect than
that coming from a single Jupiter-like object at a similar temperature. If
we were in the plane of the system, then again I think detection from the
obscuring effect may be possible.

B) The second alternative is that the initial accumulation was one of grains
and that subsequently those accumulations placed at intermediate distances
from the Sun also acquired the hydrogen and helium. The standard picture
here is of a solar nebula surrounding the Sun and flattened by rotation. The
grains (icy or silicate, depending on their distance from the Sun) again
segregate because of the gravitational field, which is now predominantly
towards the plane of the nebula, thus forming a thin grain carpet there. The
planets then grow from accumulations within this carpet.

From an observational standpoint again one would expect considerable infrared
emission from such a setup.

I realise that I am asking for more than simply the detection of an infrared
source, but any evidence would help, if only the evidence of the existence of
a flattened region surrounding a star.

Perhaps I should add that it seems unlikely that either of these situations

would persist for long once the star begins its T-Tauri phase with the
associated highly active solar wind, for then all gaseous material is simply
swept out of the system. The phenomenon we are looking for will then be
confined to young stars, perhaps less than 10^6 years old.   A general survey
would thus only have about 1 in 10^4 chance of success on age grounds alone.
The chances of success are, however, much higher if efforts are concentrated
on known young objects like those in Orion.

In conclusion, I ask you, therefore, to bear the possibility of
preplanetary systems in mind when you come to interpret your observations of
young systems and to ask you all if you think there is any chance of success
in observing such phenomenon.

References

Handbury, M.J., and Williams, I.P., 1975, Astrophys. & Sp.Sci., (in press)
McCrea, W.H., and Williams, I.P., 1965, Proc.Roy.Soc., A287, 143.
ter Haar, D., and Cameron, A.G.W., 1963, in R. Jastrow and A.G.W.Cameron
        (eds) Origin of the Solar System, Academic Press.
van de Kamp, D., 1971, Ann.Rev.Astr.Astrophys., 9, 103.
Williams, I.P., 1974, in Woszczyk and Iwanawska (eds), Exploration of the
                        planetary system, D.Reidel.
Williams, I.P., and Crampin, D.J., 1971, Mon.Not.R.astr.Soc., 152, 261.
Williams, I.P., and Cremin, A.W., 1968, Quart.J.R.astr.Soc., 9, 40.
Williams, I.P., and Handbury, M.J., 1974, Astrophys. & Sp.Sci., 30, 215.

DISCUSSION

Joseph To answer your question about whether the protoplanetary dust cloud
would be detectable in the infrared, I think F.J. Low and B.J.Smith some
years ago (1966, Nature, 212, 675) successfully interpreted the infrared
photometry of R Mon as a protoplanetary dust cloud, with the mass of dust
about equal to the mass of our planetary system.
Williams All that they showed was that the nebula could be of planetary mass.
I do not think their observations distinguished between theories, but it shows
that such a cloud can be observed. E Aur is another candidate, though not at
present in the infrared.
Martin What then distinguishes the structure of a system that is going to
produce planets from one that is not?
Williams I would have thought the primary difference is the need to form a
planar system.
Edmunds A disc shaped nebula presumably does not automatically give rise to
a planetary system, so the observation of an infrared disc is not necessarily
the observation of a protoplanetary system.
Williams I agree that it has not been proved that any disc system gives
planets. However, if I turn your comment round the other way, it is
presumably necessary to have a disc-like system (either in discrete or
continuous form) before we can get planets. It would therefore be a help to

find discs.

Wickramasinghe What do you think of the idea of dust coagulation and planet formation in Alfvenic jet streams? You would achieve much higher densities here and so your process will work much faster in these streams.

Williams I certainly think the jet stream of H. Alfven and G. Arrhenius (1970, Astrophys. Sp.Sci. 9, 3) could be an important effect within the final dust disc, though I do not know if discrete rings would be any easier to detect than a continuous carpet. However, I am not convinced that magnetic fields need to be important.

Lena I wish to mention the project considered by Labeyrie: the construction of a coronograph on an LST-type telescope should allow the detection of a Jupiter-like planet in a nearby planetary system, using the scattered stellar light in the visible. The detection of a pre-Jupiter cloud in the infrared is another matter; although the luminosity ratio with the star is favourable, the absolute value of the flux is extremely small.

Williams If most of the star's energy output is absorbed and re-emitted by the dust grains, then I would think the flux is comparable with that of the original star.

INDEX OF NAMES

INDEX OF SUBJECTS